Early Praise for *RubyMotion*

Looking for the best way to get started with RubyMotion? Read this book.

➤ **Laurent Sansonetti, lead developer of RubyMotion and founder of HipByte**

RubyMotion was a joy to read, and after zipping through the book, I found myself full of motivation to write some code! RubyMotion is a great way for longtime Ruby programmers to quickly get up to speed on many practical aspects of writing native iOS applications.

➤ **Ethan Sherbondy, creator of Listenr for iPhone**

The book is rich enough that, even though I have written many iOS apps in Objective-C and RubyMotion, with several in the App Store, I learned a few tricks. RubyMotion makes writing iOS apps faster, easier, and more enjoyable. This book gives you a jumpstart on that.

➤ **David Astels, author of *Test-Driven Development: A Practical Guide***

RubyMotion

iOS Development with Ruby

Clay Allsopp

The Pragmatic Bookshelf

Dallas, Texas • Raleigh, North Carolina

Many of the designations used by manufacturers and sellers to distinguish their products are claimed as trademarks. Where those designations appear in this book, and The Pragmatic Programmers, LLC was aware of a trademark claim, the designations have been printed in initial capital letters or in all capitals. The Pragmatic Starter Kit, The Pragmatic Programmer, Pragmatic Programming, Pragmatic Bookshelf, PragProg and the linking *g* device are trademarks of The Pragmatic Programmers, LLC.

Every precaution was taken in the preparation of this book. However, the publisher assumes no responsibility for errors or omissions, or for damages that may result from the use of information (including program listings) contained herein.

Our Pragmatic courses, workshops, and other products can help you and your team create better software and have more fun. For more information, as well as the latest Pragmatic titles, please visit us at *http://pragprog.com.*

The team that produced this book includes:

Fahmida Y. Rashid (editor)
Kim Wimpsett (copyeditor)
David J. Kelly (typesetter)
Janet Furlow (producer)
Juliet Benda (rights)
Ellie Callahan (support)

Printed in the United States of America.

ISBN-13: 978-1-937785-28-4

Printed on acid-free paper.

Book version: P3.0—July 2014

Contents

Foreword

A long time ago, Ruby replaced Perl as my favorite high-level language. I have always been a low-level programmer, writing C code on a daily basis, but I was convinced from the beginning that high-level languages such as Ruby were better suited for application development.

For a long time, application development meant desktop applications. I spent a good amount of time trying to make Ruby a first-class citizen for desktop development, working on the Ruby-GNOME, RubyCocoa, and MacRuby projects. While I can't say I achieved the success I expected, at least I tried, and it made Ruby programmers happy.

In March 2008, Apple released the first version of the iOS SDK, allowing developers to finally write custom iPhone applications in Objective-C. That day, as an Apple employee, I dreamt of the day when developers would be able to use Ruby to write iOS apps.

You are holding this book, so you know the story. I left Apple and created RubyMotion. But why on Earth would you want to use Ruby for app development?

There are many practical reasons that make Ruby a much better alternative to lower-level languages for app development, but for me, the main advantage of Ruby is that it triggers creativity.

Ruby as a language allows the creation of layered abstractions, often called *domain-specific languages*, within the language itself. This feature is definitely not unique to Ruby (I say this as a huge fan of Lisp), but Ruby programmers really seem to like creating abstractions for some reason.

Ruby has dozens of test libraries. Rails comes with several small languages defined within the framework. And just a week after RubyMotion was available, there were already several abstraction libraries for it.

Ruby programmers often call themselves *code artists* because they can feel the process of creativity when they program in Ruby. If you have programmed in Ruby, you know what I mean, because you have felt it too.

This is why Ruby matters. And this is why writing iOS apps in Ruby matters.

Laurent Sansonetti

Liege, Belgium, December 2012

Ruby on iOS

The iPhone and iOS exceeded everyone's initial expectations. In the past five years, independent developers and companies have published more than *half a million* products to the App Store that have been downloaded more than *two billion* times. But despite the huge influx of new developers and programming resources, the process of building iOS apps has remained fundamentally unchanged.

The iOS SDK was first announced in early 2008, nearly a year after the first iPhone debuted. Mac developers felt right at home since it used the same Objective-C/Xcode workflow that had existed on OS X for years. For everyone else, that day was probably the first time they heard the term *Objective-C*.

Objective-C is a robust language, but its verbosity and compiled nature are a bit out of step with the dynamic languages embraced by many of today's developers. Since Objective-C's inception in the 1980s, programmers have shifted toward Perl, PHP, Python, Ruby, and JavaScript. These "scripting" languages allowed some of the biggest websites in the world to grow and iterate with unparalleled speed by empowering flexibility and reducing complexity.

So, why haven't we seen these languages prosper on mobile yet? I mean, that is why you're here, right? The answer is that there have been no alternatives to Objective-C that allow for the trademark iOS user experience without compromising performance...well, no alternatives until now.

Hello, RubyMotion

RubyMotion (http://rubymotion.com) is such an alternative. Put simply, it allows you to develop iOS apps in Ruby without degrading the app's quality. To accomplish this, RubyMotion compiles your Ruby files to machine code; in contrast to traditional nonmobile Ruby, there's no interpreter or garbage collector to hinder performance. Your Ruby code uses the iOS SDK frameworks and classes exactly as intended by Apple, so all existing Objective-C code examples and tutorials are perfectly reusable.

Why Ruby instead of Python or some other language? For one, Ruby is already incredibly popular among web developers because of frameworks like Ruby on Rails; for new developers, coding in a familiar language means an easier transition from the Web to mobile. But most importantly, Ruby is a friendlier and more forgiving language for developers at any experience level. Whether you're a Rails veteran or just getting your feet wet with Ruby, this book will give you the foundations to create gorgeous iOS apps with no compromise in performance or developer happiness.

Why RubyMotion?

There are other alternatives to iOS development with Objective-C. HTML-based solutions like PhoneGap (http://phonegap.com) and Trigger (https://trigger.io/) are often attractive because they allow apps to be changed without additional Apple approval. However, this flexibility comes at a cost: the non-native interface elements created with HTML often create a jarring experience for users. Notably, Facebook and LinkedIn have moved away from HTML5 in their iOS apps and migrated to native versions.

Nu (https://github.com/timburks/nu) is the closest counterpart to RubyMotion: instead of using Ruby, it is a Lisp-like language you can use to write truly native iOS applications with Apple's frameworks. If you're a fan of Lisp or other functional languages, then it could be a good fit; however, RubyMotion offers more than a different language.

Unlike these other alternatives, RubyMotion is a complete tool chain that handles the entire process of creating, testing, and deploying iOS apps. Unlike the Xcode-centric Objective-C, RubyMotion development uses command-line tools such as Rake and Cocoapods (a popular iOS library manager) to increase the familiarity and ease with which developers can pick up coding for iOS. It also includes an interactive console to debug your apps and a robust, RSpec-like testing framework. No other tool or framework possesses this level of end-to-end integration for iOS development.

Reading This Book

The best programming books hit the ground running, and that's just what we'll do. Each chapter will introduce one concept and build a sample application around it. This book is intended to be read sequentially: every subsequent chapter builds on what we covered in the previous. We'll start off with the basics, such as how to draw boxes on the screen, but in just a handful of chapters we'll be interacting with an HTTP API. That means we move fast and cover just the essentials.

This isn't a reference manual for iOS development; there are many other great and extensive resources on iOS development, including Apple's official documentation. Instead, this book will get your feet just wet enough in the major topics of native iOS development so you can understand and research new information on your own.

Before we begin, you should be aware of some requirements. Since we plan on moving fast, this book assumes you're familiar with Ruby. If you haven't played around with Ruby, check out a book like *Learn to Program [Pin06]* before diving into RubyMotion. RubyMotion is currently a commercial product from HipByte; that means if you want to play, you have to pay for a license. Additionally, the RubyMotion tools work only on OS X 10.7 or newer. The iOS SDK ships with a fast desktop simulator, so you won't need a physical iOS device to test your projects on.

Online Resources

What would a modern programming book be if we didn't help you outside of the text? You should check out this book's web page (http://pragprog.com/book/carubym/rubymotion) for updates and a discussion board; you also will find all of the source code for the examples used in this book. If you're reading this book electronically, you can click the little gray box above each code sample to download it immediately.

Additionally, RubyMotion has a very vibrant community that can provide guidance. The RubyMotion Developer Center[1] has more in-depth articles on many aspects of RubyMotion that we may cover only briefly. There's also the RubyMotion user group,[2] a great place to ask specific questions and get involved.

Acknowledgments

This book was definitely not a one-man effort. First, a big thanks to the whole team at HipByte for providing support and creating the great product that is RubyMotion. I'd also like to give several high-fives to Laurent Sansonetti, Christopher Adams, David Astels, Joel Clermont, Jeff Holland, Ethan Sherbondy, Mattt Thompson, Justin de Vesine, Colin Thomas-Arnold, and Mike Clark for reviewing drafts of this book and making sure all the technical

---

1. http://www.rubymotion.com/developer-center/
2. https://groups.google.com/group/rubymotion

aspects were kosher. Additionally, I believe I owe Fahmida Y. Rashid a few cups of coffee for making this whole thing happen. And finally, a warm hug to my mom and dad for letting me do all that crazy computer stuff when I was younger.

So, what are we waiting for? Let's start making our first iOS app in Ruby!

CHAPTER 1

Creating a New App

It's time to dive into RubyMotion. We will go over what we need to install, how to create a project, and how to run and debug our apps. That might sound like a lot, but I promise you'll be surprised at just how fast we can get something up and running. Since we plan on moving fast, this book assumes you're familiar with Ruby; we won't be doing anything too tricky with the language, but you should be comfortable with the common syntax for things like blocks and instance variables.

RubyMotion 2 is a commercial product from HipByte and can be purchased from the RubyMotion website (http://rubymotion.com/). The RubyMotion tools work only on OS X 10.7 or newer, so make sure your operating system is up to date. When you purchase a license for RubyMotion, you will receive a key and a graphical installer, which will take care of everything.[1] You also need to download Xcode, Apple's IDE, from the Mac App Store. Installing Xcode also installs some developer tools RubyMotion relies on (such as the iOS Simulator), but you don't have to work on RubyMotion projects using it or any other IDE.

Instead, RubyMotion uses command-line tools, and you can use any text editor you want. There are add-ons for many popular editors that help with things like code completion and build integration.[2] RubyMotion also builds on top of existing Ruby tools such as RubyGems and Rake, so Ruby developers will feel right at home.

Once everything is installed, you're ready to take the dive. Read on!

1. For detailed installation instructions, visit the RubyMotion documentation at http://www.rubymotion.com/developer-center/guides/getting-started/.
2. Check out the RubyMotion documentation at http://www.rubymotion.com/developer-center/articles/editors/ for a short list.

1.1 Creating a New App

RubyMotion exists on the command line, so you should know the basics of navigating around a shell. The motion command is installed when you run the RubyMotion installer. It's equivalent to the rails command in the Ruby realm. motion manages individual projects in addition to the core RubyMotion tools. For example, motion update automatically updates RubyMotion and motion support lets you file bug reports.

Let's open a terminal and navigate to where you want to create your Ruby-Motion projects. We're going to create a few projects over the course of this book, so your work area should preferably be somewhere without any other files or folders. Once you've picked that out, run motion create HelloMotion.

```
$ motion create HelloMotion
Create HelloMotion
Create HelloMotion/.gitignore
Create HelloMotion/app/app_delegate.rb
Create HelloMotion/Gemfile
Create HelloMotion/Rakefile
Create HelloMotion/resources/Default-568h@2x.png
Create HelloMotion/spec/main_spec.rb
```

motion create makes a HelloMotion folder in the current directory and adds some more files inside. These generated files and folders form the essential skeleton of a RubyMotion project. Go ahead and cd into it (cd ./HelloMotion) so that we can take a look. You'll run all of our subsequent RubyMotion commands from within this folder, so definitely keep a terminal window or tab open pointing to the directory.

For now, we'll just talk about two of the files it created: Rakefile and ./app/app_delegate.rb.

The Rakefile is created at the root of a project. It's where we configure apps and include settings such as the app's name, icon, resources, and source code locations. Before we get too far ahead of ourselves, let's start with the basics.

The Rakefile gets its name from the rake command. rake is a command that allows you to execute arbitrary Ruby code via the command line. These chains (known as *tasks*) are loaded from the directory's Rakefile and become executable in the form rake {task name}. For example, RubyMotion uses the rake device tasks to build an app to a device.

As of RubyMotion 2.28, the Rakefile will be generated to look like this:

```
first_app/HelloMotion/Rakefile
# -*- coding: utf-8 -*-
$:.unshift("/Library/RubyMotion/lib")
require 'motion/project/template/ios'
begin
  require 'bundler'
  Bundler.require
rescue LoadError
end
Motion::Project::App.setup do |app|
  # Use `rake config' to see complete project settings.
  app.name = 'HelloMotion'
end
```

The first line specifies the the the file encoding; the next two lines import necessary RubyMotion libraries into our project. The code between the following begin() and end() block allows our project to use Bundler (http://bundler.io/) to manage Ruby dependencies. RubyMotion creates a Gemfile in our project by default to use with Bundler, but we won't be using Bundler in this book. Next, we call Motion::Project::App.setup. This is what actually sets up the Rakefile to build a RubyMotion app. We pass it a block where we configure the app object. By default, RubyMotion sets the app's name to correspond to what we passed in motion create.

Earlier we mentioned rake device, but that's nowhere in the Rakefile. It and the other RubyMotion tasks are defined when we require() motion/project. To list all the tasks, let's run rake --tasks in a terminal.

```
$ rake --tasks
rake archive                   # Create an .ipa archive
rake archive:distribution      # Create an .ipa archive for AppStore
rake build                     # Build everything
rake build:device              # Build the device version
rake build:simulator           # Build the simulator version
rake clean                     # Clear local build objects
rake clean:all                 # Clean all build objects
rake config                    # Show project config
rake crashlog
rake crashlog:device
rake ctags                     # Generate ctags
rake default                   # Build the project, then run the simulator
rake device                    # Deploy on the device
rake profile                   # Same as profile:simulator
rake profile:device            # Run a build on the device through Instruments
rake profile:simulator         # Run a build on the simulator through Instruments
rake simulator                 # Run the simulator
rake spec                      # Same as 'spec:simulator'
rake spec:device               # Run the test/spec suite on the device
rake spec:simulator            # Run the test/spec suite on the simulator
rake static                    # Create a .a static library
```

> **$:.unshift Who?**
>
> If you're not intimately familiar with Ruby, the first thing you may think when reading the Rakefile is "Wait...$:.unshift()?" Strange-looking indeed. What that line does is tell Ruby this: "When we use require(), also look in the /Library/RubyMotion/lib directory to search for what we're requiring." motion/project/template/ios resides there, and without that initial $:.unshift(), no RubyMotion code would be found!

There are lots of neat toys for us to play with. What isn't evident is that the default, plain rake command is configured to be the same as building and running the app on the simulator.

Let's take a look at the simulator by running rake in a terminal.

```
$ rake

   Build ./build/iPhoneSimulator-7.1-Development
 Compile ./app/app_delegate.rb
  Create ./build/iPhoneSimulator-7.1-Development/HelloMotion.app
    Link ./build/iPhoneSimulator-7.1-Development/HelloMotion.app/HelloMotion
  Create ./build/iPhoneSimulator-7.1-Development/HelloMotion.app/Info.plist
  Create ./build/iPhoneSimulator-7.1-Development/HelloMotion.app/PkgInfo
  Create ./build/iPhoneSimulator-7.1-Development/HelloMotion.dSYM
Simulate ./build/iPhoneSimulator-7.1-Development/HelloMotion.app
(main)> _
```

Give it a second to finish its business and...you'll see an iPhone simulator pop up with our app running. Congratulations! You've just created your first iOS app in Ruby.

Additionally, your terminal should be displaying a new prompt. We can enter new Ruby code in the terminal and watch it execute on the fly; this is just like the rails console and irb commands in the Ruby and Rails world. Altogether, your screen should look like the one in Figure 1, *Prompt with iPhone simulator*, on page 5.

Hooray! But how did that really happen? How did we get from app.name = 'HelloMotion' to an iPhone emulator popping up? To find the answer, we must leave rake and explore the app's actual code.

1.2 Where It Begins: AppDelegate

Before we even setup() the app in the Rakefile, our app comes with some sensible defaults. Most importantly, it automatically compiles all Ruby files in the app directory of our project. Remember that app_delegate.rb we mentioned earlier?

Figure 1—Prompt with iPhone simulator

It so happens that code was compiled with the app. In fact, it was the *only* code compiled with our app, so it's probably a good idea to see what it does.

first_app/HelloMotion/app/app_delegate.rb
```ruby
class AppDelegate
  def application(application, didFinishLaunchingWithOptions:launchOptions)
    true
  end
end
```

Hmm, all that did was define an AppDelegate class with one method. There's not even a superclass, so how does this do anything at all?

Well, there's a comment in our Rakefile on page 3 that tells us to run the rake config command to see all project settings. Let's run that.

```
$ rake config
background_modes    : []
build_dir           : "./build"
..                    ..
delegate_class      : "AppDelegate"
..                    ..
xcode_dir           : "/Applications/Xcode.app/Contents/Developer"
```

There's that AppDelegate class, and it's assigned as the delegate_class. What's that do? Well, RubyMotion actually looks for a class with the name we assign as delegate_class and uses that as the *application delegate* when launching our app. We could've called our class SuperAppDelegate and changed our Rakefile to have the same result:

```
Motion::Project::App.setup do |app|
  app.name = 'HelloMotion'
  app.delegate_class = "SuperAppDelegate"
end
```

What exactly is the application delegate? When the user launches our app, the system does some setup work to make sure everything loads correctly. We need to give the system an object that can take over after that setup is done; we refer to that object as the application delegate. It gets callbacks for all sorts of application-level events, such as when the app starts, ends, goes to the background, gets a push notification, and does all that other good stuff.

In the code generated by motion create, AppDelegate implements only the method application:didFinishLaunchingWithOptions:. This method and notation look a little different from normal Ruby code because of the named argument didFinishLaunchingWithOptions: shoved in the middle. For more information about named arguments, check out *Why Are There Colons in My Ruby?*, on page 6.

Why Are There Colons in My Ruby?

In many languages, functions look like this: obj.makeBox(origin, size). This can be a bit of a pain because now you need to look up the implementation of that function and figure out the order of the arguments. Objective-C uses *named parameters* to make this easier. In Objective-C, that same function looks like this: [obj makeBoxWithOrigin: origin andSize: size]. See how each variable directly follows the part of the function name that refers to it, removing ambiguity? Pretty clever. We refer to those functions by putting colons in place of variable names, like so: makeBoxWithOrigin:andSize:.

Historically, named arguments don't exist in Ruby;[a] instead, it has traditional methods like obj.make_box(origin, size). RubyMotion adds named arguments to its implementation of Ruby so the Objective-C APIs are compatible, like so: obj.makeBox(origin, andSize: size). That andSize isn't just an extra variable; it's a real part of the method name. obj.call_method(var1, withParam: var2) is totally different from obj.call_method(var1, withMagic: var2), despite having the same obj.call_method() form. Also note that named arguments are different from passing a Hash as the lone argument, which is common in Ruby libraries.

---

a.   Named arguments like those in RubyMotion are available as of Ruby 2.0.

application:didFinishLaunchingWithOptions: is called when the system finishes setting up the app and is ready for us to do our own setup. For now, just assume it will always return true. Some apps may want to use the launchOptions argument to determine whether the app should be started, but most of the time you won't need that.

But aside from returning true and allowing the app to launch, our AppDelegate doesn't do anything. Let's change that by showing a quick UIAlertView.

1.3 Showing a Message with UIAlertView

Now that we understand how our app boots up and we have our entry point, let's try something. We're going to show a UIAlertView, which is the standard blue pop-up you see when you get an error message or push notifications pre-iOS5. Let's change application:didFinishLaunchingWithOptions: to add an alert right before returning.

first_app/HelloMotion_alert/app/app_delegate.rb
```ruby
class AppDelegate
  def application(application, didFinishLaunchingWithOptions:launchOptions)
➤    @alert =
➤      UIAlertView.alloc.initWithTitle("Hello",
➤        message: "Hello, RubyMotion",
➤        delegate: nil,
➤        cancelButtonTitle: "OK",
➤        otherButtonTitles: nil)
➤    @alert.show
➤
➤    puts "Hello from the console!"

    true
  end
end
```

We create our alert using UIAlertView.alloc.initWith..., which is a prime example of the verbose Objective-C syntax used in the iOS SDK. The initializer here takes many arguments, but the only ones we care about are title, message, and cancelButtonTitle. Once those are set, all we need to do is call show(), and the alert will appear.

We also added a call to the logging statement puts(). You can pass it a normal string to print verbatim, or you can pass it any normal object that you want some information about. This will be the output visible in the debugging console while our app runs.

Speaking of which, let's run rake one more time and check out our blue pop-up! (See Figure 2, *A UIAlertView in your app*, on page 8.)

Figure 2—A **UIAlertView** in your app

In your terminal window, you should see (main)> Hello from the console! output in the emulator. While we're looking here, let's take a deeper look into the RubyMotion debugger.

1.4 Interactive Debugging

Earlier we mentioned that RubyMotion ships with an interactive debugger. This is what appeared in our terminal after running our first app in Section 1.2, *Where It Begins: AppDelegate*, on page 4. Any code entered in the debugger is executed line by line, and the changes you make are reflected immediately in the simulator.

The debug console can be a really powerful tool not only for debugging but also for rapid prototyping. We can try different fonts and colors or play with the position of UI elements without recompiling the app each time. But all that versatility may make it seem daunting to actually figure out how to use

All About Initializers: new, alloc, init, and More

Objects in Ruby are usually created using the static new() method, like Array.new, and then initialization steps are performed in the initialize() instance method. However, you may have noticed that we created our UIAlertView using UIAlertView.alloc.initWithTitle...—what's up with that?

In Objective-C, alloc() is the equivalent of new() and init() is the equivalent of initialize(). The convention is to use many different initWith...() methods for specific behaviors, hence our use of initWithTitle:message:...().

When you use one of those lengthier initializers in RubyMotion, you must use the alloc.initWith... syntax; however, if all you need to do is use the plain init initializer, as in Class.alloc.init, then calling Class.new() will do the same thing.

the console effectively. Let's play with our app in the console and see if we can have some fun.

Let's rake our app one more time. After compilation and setup, the terminal should display a prompt while the simulator is open. Because of our puts() in AppDelegate, the Hello from the console! should be above the highlighted prompt. Don't close the UIAlertView yet because we're going to play with it.

Let's find the instance of our AppDelegate. Remember that it's set as our application delegate, so first we probably need to grab information about our application. Luckily, the UIApplication class represents an application, and we access our application's instance using UIApplication.sharedApplication().

```
(main)> app = UIApplication.sharedApplication
=> #<UIApplication>
```

Care to take a guess how we get our AppDelegate out of that? Probably a delegate() method, right?

```
(main)> delegate = app.delegate
=> #<AppDelegate>
```

Excellent; now we can grab our UIAlertView using Ruby's instance_variable_get(). We can change the alert's message and title and run the commands in the relevant code:

```
(main)> alert = delegate.instance_variable_get("@alert")
=> #<UIAlertView>
(main)> alert.title = "Goodbye"
=> "Goodbye"
(main)> alert.message = "You say yes?"
=> "You say yes?"
(main)> alert.show
```

The results show immediately:

Finally, we can programmatically hide our alert with alert.dismiss. Even though we used only this one alert object, we were able to interact with and change our app's UI with a few easy commands.

See, getting up and running with RubyMotion wasn't so terrible. With just motion create and rake, we started a new project and were able to run and debug our app. And if you ask me, that's a sight easier than booting up Xcode and going through its many project creation wizards and configuration screens. Plus, the RubyMotion debugging workflow is miles more intuitive than using the low-level gdb and lldb tools in Xcode. In all, RubyMotion makes for a buttery-smooth transition into mobile development rather than jumping into the jarring world of Xcode and Objective-C.

But this is just the start: we need to put elements that are more interesting than blue alerts on the screen. In the next chapter, *Filling the Screen with Views*, we'll cover how to place and animate basic boxes, buttons, and labels in our app.

CHAPTER 2

Filling the Screen with Views

As shown by everything from reviews to lawsuits, iOS apps are recognized for their visual design. Designers can craft the most beautiful (or crazy) graphics for an app, but it's the developer's job to bring them to life. But before we can make apps like Square, we need to start with the basics: putting simple shapes and objects on the screen.

2.1 All About Views and UIView

All of the "stuff" you see and interact with in an app are called *views*. Each distinct screen is made of dozens of views, from tiny icons to buttons and text entry fields. The base view of an app is called the *window*, and all other views exist within it. Apps can even have multiple windows, but that's a little bit beyond our scope.

In code, views are UIView objects; all graphics on the screen are descendants of UIView. The window is a subclass of UIView called UIWindow. When our app starts, we set up a window and then begin adding new views to it. These additional views are known as *subviews*.

Subviews are visually stacked on top of each other within the parent view. Every view, not just the window, can have subviews. When you move a view, you also move its subviews. Programmatically, UIViews use a subviews property. This returns an array of the subviews sorted by back-to-front visibility (so a subview at index 0 is behind a subview at index 1).

Views have a rectangular shape and 2D location coordinates. The *origin* location (0, 0) is the top-left corner of a view, so its bottom-right point corresponds to (view's width, view's height). In code, UIView uses the frame property to describe these properties. Every frame has an origin with which you can access the x and y coordinates, as well as a size property that defines the view's width and height.

That's a lot to take in, so for help refer to the following:

We can do a whole lot using just subviews and frame. Eventually we want to build the next GarageBand, but first we need to start with a lone colored box.

2.2 Making Shapes and Colors

Let's get to some code. Create a new app called Boxey (remember how? motion create Boxey). Now we need to add a window to our app.

Adding a UIWindow should be one of the first things your app does, so we need to add it in AppDelegate's application:didFinishLaunchingWithOptions:, like this:

```
views/Boxey_shape/app/app_delegate.rb
class AppDelegate
  def application(application, didFinishLaunchingWithOptions:launchOptions)
    @window = UIWindow.alloc.initWithFrame(UIScreen.mainScreen.bounds)
    @window.backgroundColor = UIColor.whiteColor
    @window.makeKeyAndVisible
```

There are a couple of new classes in there, so let's walk through it. We created our UIWindow using initWithFrame:. We use this method to create all UIViews, not just windows. It takes a CGRect object, but for windows we usually want a frame that fills the display. This is why we use UIScreen instead of specifying our own size.

UIScreen is an object that contains information about the displays our app is running on (chiefly the mainScreen). Its bounds() returns the CGRect that describes the visual size of our application. Apple's design guidelines suggest that your application should fill the entire screen (including underneath the status bar) and bounds() will give us that dimension.

Next we set the background color with a UIColor. It has some obvious defaults such as whiteColor() and blueColor(), but we can also create arbitrary hues. UIWindow has a black background by default, and it will make our exercises a bit simpler if we set it to white.

So, we create our window with the appropriate settings and call makeKeyAndVisible(). This is a special method for UIWindow that tells the system it will be the window receiving touch events and should be drawn to the screen.

That's a lot of explanation, I know, but thankfully setting up a window needs to be done only once. More exciting is how we continue application:didFinishLaunching: by adding our first new UIView.

views/Boxey_shape/app/app_delegate.rb
```
    @blue_view = UIView.alloc.initWithFrame(CGRect.new([10, 40], [100, 100]))
    @blue_view.backgroundColor = UIColor.blueColor
    @window.addSubview(@blue_view)

    true
  end
end
```

Again, we use initWithFrame:, except now we create a new CGRect object with the initializer new([x, y], [width, height]). The two arrays correspond to the origin and size subproperties of CGRect, as in rect.origin.x and rect.size.width. In some code, you might see CGRects created using the CGRectMake(x, y, width, height) method. RubyMotion also supports this style, which is how it works in the Objective-C iOS SDK, but CGRect.new is more idiomatic in Ruby. Remember that UIWindow is a subclass of UIView, so we then set its background color just like before.

Lastly we use addSubview: to add the view to our window. Even though we can access the subviews() property of UIView, we can't append or insert to it directly. Instead, we use specific methods like addSubview for those tasks.

Go ahead and run our app (just rake, remember?) and observe our blue box. Even though it looks like the following, at least we have something on the screen!

We'll start to make it more useful with buttons in Section 2.3, *Adding Interaction with UIButton*, on page 14.

Don't Add Views to Windows

Adding lone UIViews directly to the window is frowned upon and generally not a great idea, but it's a nice way to learn the basics. It's OK while you're still learning, but don't do it in production code. We'll get to the proper way of adding views in the next chapter, *Organizing Apps with Controllers*.

2.3 Adding Interaction with UIButton

The iOS APIs include many standard UIView subclasses. Some add intricate functionality, while others are merely ornamental. UIButton is one such view, and it does exactly what it sounds like. Let's add it to our app and see how it works.

UIButtons are created differently than normal UIViews. Instead of using alloc.initWithFrame, UIButtons are initialized using the static method UIButton.buttonWithType:. We pass an integer option, usually UIButtonTypeSystem or UIButtonTypeCustom, and get a new button. Many iOS APIs use verbose integer constants like these because more idiomatic Ruby Symbols don't exist in the original Objective-C. Anyhow, take a look at how we add our button in AppDelegate.

```
views/Boxey_buttons/app/app_delegate.rb
@window.addSubview(@blue_view)
@add_button = UIButton.buttonWithType(UIButtonTypeSystem)
@add_button.setTitle("Add", forState:UIControlStateNormal)
@add_button.sizeToFit
@add_button.frame = CGRect.new(
  [10, @window.frame.size.height - 10 - @add_button.frame.size.height],
  @add_button.frame.size)
@window.addSubview(@add_button)
```

UIButtonTypeSystem gives us a transparent button with blue text. Usually apps will customize their button's image or backgroundImage with custom designs, but the system button doesn't look bad either. Next we assign the button a title using setTitle:forState:; the title part is obvious, but what the heck is a control state?

UIButton is actually a subclass of UIControl, which is a direct subclass of UIView. UIControl provides common behaviors for interactive elements like buttons and switches, and each control has a state that describes how it can be interacted with. Depending on how we want the user to interact with them, controls can have states like UIControlStateNormal or UIControlStateDisabled.[1] UIButton uses the state

---

1. You can find a complete list of control states at http://developer.apple.com/library/ios/#documentation/uikit/reference/UIControl_Class/Reference/Reference.html.

it inherited from UIControl to change the text depending on its current state. Many control elements follow this pattern, so it's good to be familiar with it.

After setting the title, we call sizeToFit(). This is another method every UIView has, but how it works will change from subclass to subclass. In UIButton's case, it will resize itself to fit its title with appropriate padding.

After that, we set the frame with some customized alignments and add the button as another subview to the window. We rake again and see our understated button at the bottom of the screen. Unfortunately, it doesn't do more than slightly change color when you tap it:

UIButtonTypeSystem   UIButtonTypeSystem

Naturally, there's a way to change that.

UIControls allow us to intercept events such as touching and dragging using addTarget:action:forControlEvents:. This method will call the action method on the target object when the control event(s) occur; again, this un-Ruby pattern is another component rooted in the original Objective-C APIs. The event passed in this method is actually a *bitmask* of UIControlEvents,[2] so if you want to register the action for multiple events, it looks like this: forControlEvents: (UIControlEventTouchDown | UIControlEventTouchUpInside).

When our button is tapped, we'll add a new blue box to the screen, like so:

```
views/Boxey_buttons/app/app_delegate.rb
  @window.addSubview(@add_button)
  @add_button.addTarget(
    self, action:"add_tapped", forControlEvents:UIControlEventTouchUpInside)
  true
end

def add_tapped
  new_view = UIView.alloc.initWithFrame(CGRect.new([0, 0], [100, 100]))
  new_view.backgroundColor = UIColor.blueColor
  last_view = @window.subviews[0]
  new_view.frame = CGRect.new(
    [last_view.frame.origin.x,
      last_view.frame.origin.y + last_view.frame.size.height + 10],
    last_view.frame.size)
  @window.insertSubview(new_view, atIndex:0)
end
```

---

2. Find a complete list of control events at http://developer.apple.com/library/ios/#a/uikit/reference/ UIControl_Class/Reference/Reference.html#//apple_ref/occ/cl/UIControl.

Aside from the button callback, we've talked about all these methods before. UIControlEventTouchUpInside triggers when the user lifts their *touch up*, *inside* of the button's frame, so it acts as an event to track taps. You can decide to be clever and arrange the boxes into multiple columns, but I'll leave the details of the implementation up to you. Here, run our app, hit the button a few times, and admire our fine work (Figure 3, *Tapping to add a box*, on page 16).

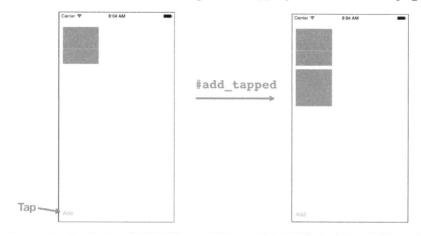

Figure 3—Tapping to add a box

2.4 Animating Views

We talked about how important views are on iOS, but apps are also known for their smooth animations. Perhaps one reason slick transitions and niceties are so pervasive in iOS apps is that they're so darn easy to implement. And we're going to add that to our little box app right now.

Let's add another button, Remove, which will fade out the most recently added view and slide all others to new positions in its place. That might sound complicated, and on other platforms or frameworks it might be, but the iOS animation APIs make it painless. All we do is tell the system what properties of our views to animate and how long that animation should take.

We will add yet another button to our AppDelegate and wire its target/action callbacks for removing a view.

```
views/Boxey_animations/app/app_delegate.rb
@add_button.addTarget(
  self, action:"add_tapped", forControlEvents:UIControlEventTouchUpInside)
@remove_button = UIButton.buttonWithType(UIButtonTypeSystem)
@remove_button.setTitle("Remove", forState:UIControlStateNormal)
@remove_button.sizeToFit
```

```
@remove_button.frame = CGRect.new(
  [@add_button.frame.origin.x + @add_button.frame.size.width + 10,
    @add_button.frame.origin.y],
  @remove_button.frame.size)
@window.addSubview(@remove_button)
@remove_button.addTarget(
  self, action:"remove_tapped",
  forControlEvents:UIControlEventTouchUpInside)
```

Pretty easy, right? Basically, all we did was set a frame and add a subview; that's nothing new. Now we need to implement that remove_tapped() callback. It's going to be longer than our add_tapped() method, so we'll take it slow. First, we need to find the objects we're interested in.

views/Boxey_animations/app/app_delegate.rb
```
def remove_tapped
  other_views = @window.subviews.reject { |view|
    view.is_a?(UIButton)
  }
  last_view = other_views.last
  return unless last_view && other_views.count > 1
```

Because our buttons are also subviews of the window, we need to prune them and make sure we deal only with the blue boxes. There are better ways to architect this (such as storing the boxes in some independent array), but this works with what we have. Next, we do the actual animations!

views/Boxey_animations/app/app_delegate.rb
```
  animations_block = lambda {
    last_view.alpha = 0
    last_view.backgroundColor = UIColor.redColor
    other_views.reject { |view|
      view == last_view
    }.each { |view|
      new_origin = [
        view.frame.origin.x,
        view.frame.origin.y - (last_view.frame.size.height + 10)
      ]
      view.frame = CGRect.new(
        new_origin,
        view.frame.size)
    }
  }
  completion_block = lambda { |finished|
    last_view.removeFromSuperview
  }
  UIView.animateWithDuration(0.5,
    animations: animations_block,
    completion: completion_block)
end
```

Animations revolve around the UIView.animateWithDuration:animations: group of methods (you can also use animateWithDuration:delay:options:animations:completions: if you need to fine-tune things). Any alterations to your views made in the lambda we pass for animations will animate if possible. Beyond the basics like frame and opacity, most sensible properties of UIView will work as expected.[3] In our case, we fade out the box by setting the floating-point alpha to zero. Then we enumerate through all the other views and adjust their frames.

We use the optional completion: argument to get a callback when the animation finishes. This block takes one boolean argument, which tells us if the callback has been fired when the animation actually completed (the callback may fire prematurely if the animation has been canceled elsewhere). This is a good place to clean up our views, which we do here by invoking removeFromSuperview(). This will remove the view from its parent's subviews and be erased from the screen.

The animation function looks a bit strange because of the multiple lambdas, but it's no different from changing those properties of a view when they're static. Give it a rake, add some boxes, and then watch them float away with the Remove button:

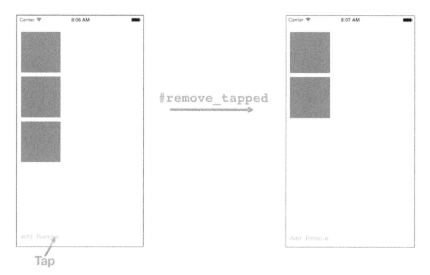

2.5 Adding Static Text with UILabel

Animations are fun, but we also need to display information long enough for the user to actually read it. In most cases, we can use UILabel to display static

---

3. You can find a full list of animatable properties at:
   http://developer.apple.com/library/ios/#documentation/uikit/reference/uiview_class/uiview/uiview.html.

text. Labels can be very flexible, allowing you to change everything from the font to minute adjustments with the text baseline, and are really easy to get up and running. Let's add one to our little app.

We're going to add a UILabel to each box, displaying its index in subviews. We probably wouldn't ship that sort of feature, but it's really handy for debugging and might give us a better idea of what's going on in our animation. UILabel is really lightweight, so it won't be a pain to add.

Adding labels is going to occur in a new method called add_label_to_box(). This method will figure out the index of a given box's UIView instance and add the correct UILabel. This is the important part of the code, so let's take a look at it first.

```ruby
views/Boxey_label/app/app_delegate.rb
def add_label_to_box(box)
  box.subviews.each do |subview|
    subview.removeFromSuperview
  end

  index_of_box = @window.subviews.index(box)
  label = UILabel.alloc.initWithFrame(CGRectZero)
  label.text = "#{index_of_box}"
  label.textColor = UIColor.whiteColor
  label.backgroundColor = UIColor.clearColor
  label.sizeToFit
  label.center = [box.frame.size.width / 2, box.frame.size.height / 2]
  box.addSubview(label)
end
```

We start by removing all subviews from box, which handles the case where we call this method multiple times on the same view (which we will). Our label is initialized with CGRectZero, which is shorthand for a rectangle at the origin and no size. After we set the text appropriately, we call sizeToFit() just like UIButton. The UILabel implementation of sizeToFit() will precisely fill the frame to fit the text, leaving no padding. Then we use the center property of UIView, which is shorthand for putting the center of a view at a point (as opposed to the upper-left corner).

Remember how we said subviews are positioned within their parent? Even though we set the label to be centered at a coordinate like (50, 50), it can exist at a different point within the window. As our animation slides the box, its label will move too.

Not too bad, right? The only UILabel-exclusive properties in this example are text and textColor; everything else is inherited from UIView. Now we need to actually call this method.

We usually want to go through all the boxes each time we update the labels so we can be absolutely sure our labels are in sync with subviews. To make our lives easier, we're going to refactor the logic for picking out boxes from @window.subviews into one method that simply returns only the boxes.

views/Boxey_label/app/app_delegate.rb
```ruby
def boxes
  @window.subviews.reject do |view|
    view.is_a?(UIButton) or view.is_a?(UILabel)
  end
end
```

We can combine our two new methods into one really great helper method that takes care of everything.

views/Boxey_label/app/app_delegate.rb
```ruby
def add_labels_to_boxes
  self.boxes.each do |box|
    add_label_to_box(box)
  end
end
```

Finally, we can put these to some use, first in application:didFinishLaunchingWithOptions:

views/Boxey_label/app/app_delegate.rb
```ruby
@window.addSubview(@blue_view)
add_labels_to_boxes
```

and then down in add_tapped().

views/Boxey_label/app/app_delegate.rb
```ruby
@window.insertSubview(new_view, atIndex:0)
add_labels_to_boxes
```

Lastly, we need to reset the labels for each box after we run the removal animation. We're going to change our other_views to use self.boxes instead of its own array construction. Then we're going to use our handy add_labels_to_boxes() to sync all the labels again.

views/Boxey_label/app/app_delegate.rb
```ruby
def remove_tapped
  other_views = self.boxes
  last_view = other_views.last
```

views/Boxey_label/app/app_delegate.rb
```ruby
completion_block = lambda { |finished|
  last_view.removeFromSuperview
  add_labels_to_boxes
}
```

Whew. Our UILabel was only a small part of our changes, but now we can clearly see how our view hierarchy behaves at runtime. Run the app, and you should see labels appear as in the following figure.

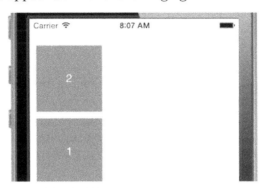

Figure 4—Labels reset for each box

2.6 Making Text Dynamic with UITextField

Most apps have more than just buttons and labels; usually we need the user to enter some data, like a tweet or email address. UITextField is the basic view we use to grab string-type input, and now we're going to add one to our little box app. It will let the user pick the UIColor of our boxes using simple commands like "red" and "blue." Let's get started.

First we need to add the field to the view hierarchy. UITextField puts many configuration options at our disposal, from fonts to the look of the Return key. We won't be using all of its properties today, but you can consult Apple's documentation on the class for more information.[4] Our field should be added in app_delegate.rb like so:

```
views/Boxey_textfield/app/app_delegate.rb
@remove_button.addTarget(
  self, action:"remove_tapped",
  forControlEvents:UIControlEventTouchUpInside)

@color_field = UITextField.alloc.initWithFrame(CGRectZero)
@color_field.borderStyle = UITextBorderStyleRoundedRect
@color_field.text = "Blue"
@color_field.enablesReturnKeyAutomatically = true
@color_field.returnKeyType = UIReturnKeyDone
@color_field.autocapitalizationType = UITextAutocapitalizationTypeNone
```

4. You can find the UITextField reference at http://developer.apple.com/library/ios/#documentation/uikit/reference/UITextField_Class/Reference/UITextField.html.

```
@color_field.sizeToFit
@color_field.frame = CGRect.new(
  [@blue_view.frame.origin.x + @blue_view.frame.size.width + 10,
    @blue_view.frame.origin.y + @color_field.frame.size.height],
  @color_field.frame.size)
@window.addSubview(@color_field)
```

> `@color_field.delegate = self`

Like every other view in the window, we spend some time setting up the frame and positioning it exactly where we want it. returnKeyType() and similar properties control exactly what they say; the only cryptic property we use is UITextBorder-StyleRoundedRect, which adds a nice border and inner shadow to our field. By default, UITextFields have empty backgrounds and no default styling.

The most important part of our addition is setting @color_field's delegate. Much like our application uses AppDelegate as its delegate, other objects use the delegation pattern as a way of sending callback events. The UITextFieldDelegate specification lists all of the methods the delegate object can implement.[5] We aren't required to implement any of them, but we will implement textFieldShould-Return: to intercept when the Return/Done key is pressed.

views/Boxey_textfield/app/app_delegate.rb
```
def textFieldShouldReturn(textField)
  color_tapped
  textField.resignFirstResponder
  false
end
```

resignFirstResponder looks a little cryptic, but in reality it simply hides the keyboard. In iOS, there's a concept of a *responder chain* that determines how events such as taps are propagated among our objects. We won't deal with the responder chain in this book, but the important thing to remember is that the first responder of a text field is almost always the virtual keyboard. You also may notice that we explicitly return false from textFieldShouldReturn:, but why is that? Whatever you return from this method decides whether the UITextField carries out the default behavior of its Return key; in our case, we're hiding the keyboard, and the normal action should be avoided.

Finally, we need to implement the color_tapped() method we called in textFieldShould-Return:. We're going to read the text property of the text field and use the Ruby metaprogramming send() method to create a UIColor from that string.

---

5. You can find the UITextFieldDelegate reference at http://developer.apple.com/library/ios/#Documentation/ UIKit/Reference/UITextFieldDelegate_Protocol/UITextFieldDelegate/UITextFieldDelegate.html.

```
views/Boxey_textfield/app/app_delegate.rb
def color_tapped
  color_prefix = @color_field.text
  color_method = "#{color_prefix.downcase}Color"
  if UIColor.respond_to?(color_method)
    @box_color = UIColor.send(color_method)
    self.boxes.each do |box|
      box.backgroundColor = @box_color
    end
  else
    UIAlertView.alloc.initWithTitle("Invalid Color",
        message: "#{color_prefix} is not a valid color",
        delegate: nil,
        cancelButtonTitle: "OK",
        otherButtonTitles: nil).show
  end
end
```

Since we're feeling friendly, we alert the user if there is no such UIColor for their input. But if we do succeed in creating a color object, we assign it to a @box_color instance variable. We need to go back to other parts of the code to make sure they also use @box_color; that way, events like adding a new box work as expected.

```
views/Boxey_textfield/app/app_delegate.rb
@window.makeKeyAndVisible
@box_color = UIColor.blueColor
@blue_view = UIView.alloc.initWithFrame(CGRect.new([10, 40], [100, 100]))
@blue_view.backgroundColor = @box_color
@window.addSubview(@blue_view)
```

```
views/Boxey_textfield/app/app_delegate.rb
def add_tapped
  new_view = UIView.alloc.initWithFrame(CGRect.new([0, 0], [100, 100]))
  new_view.backgroundColor = @box_color
```

All we're doing here is changing the hard-coded use of blueColor to our new instance variable. And there's one more thing: we need to fix our boxes() method to ignore the new UITextField.

```
views/Boxey_textfield/app/app_delegate.rb
def boxes
  @window.subviews.reject do |view|
    view.is_a?(UIButton) or view.is_a?(UILabel) or view.is_a?(UITextField)
  end
end
```

Fantastic; let's run rake again and play with the text field (see Figure 5, *Playing with the text field*, on page 24). Be sure to try more exotic colors like "cyan" and "magenta," too.

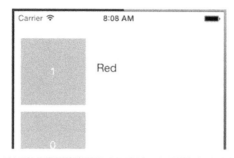

Figure 5—Playing with the text field

2.7 Exploring RubyMotion Libraries

We whipped up a pretty interesting app using a relatively small amount of code; however, we also used some vestigial Objective-C patterns that look obviously out of place. This is one area where the RubyMotion community is stepping up and wrapping un-Ruby code into more idiomatic structures. Several libraries and RubyGems[6] are available that could have helped us manage our views.

For example, Sugarcube (https://github.com/rubymotion/sugarcube) would have allowed us to replace those long animation method names with very concise functions such as fade_out() and move_to().

```
last_view.fade_out { |view|
  last_view.removeFromSuperview
}

other_views.each do |view|
  new_origin = [
    view.frame.origin.x,
    view.frame.origin.y - (last_view.frame.size.height)
  ]

  view.move_to new_origin
end
```

Much better, right? And for more complex apps, the Teacup library[7] allows you to construct views using CSS-esque style sheets. Our blue boxes might have Teacup style sheets defined like this:

---

6. For a full explanation of RubyGems and RubyMotion, check out *Third-Party Libraries and RubyMotion*, on page 49.
7. https://github.com/rubymotion/teacup

```
Teacup::Stylesheet.new :app do
  style :blue_box,
    backgroundColor: UIColor.blueColor,
    width: 100,
    height: 100
end
```

Third-party libraries like these are helping RubyMotion become more than just an Objective-C/Ruby mashup. As you can see in the previous examples, they can dramatically change how we express what we are trying to accomplish in code. Later in *Representing Data with Models* and *Example: Writing an API-Driven App*, we'll actually use some third-party RubyMotion frameworks to simplify otherwise complex elements of our apps.

But even without those niceties, we've gone from an empty app to an interactive, animated UI in the span of a few quick examples. There are far more included UIView subclasses than we have time for, but the ones we've covered should make those a cinch to learn when the time comes.

By making a slightly more ambitious app, we have also gotten a chance to see how the Ruby language can make our iOS development lives a little easier. Take my word for it, RubyMotion methods such as Array#select and string formatting ("#{ruby_code_here}") are more concise than their Objective-C counterparts. Then again, we still have some very non-Ruby practices that leave much to be desired, such as UITextField's delegate pattern.

In the course of working on views, our AppDelegate got pretty crowded with all kinds of code: helper functions, button callbacks, view creation...the works. Real apps have much more robust organization in the form of controllers, which we'll cover now in *Organizing Apps with Controllers*.

CHAPTER 3

Organizing Apps with Controllers

iOS apps usually consist of more than simple rectangles and buttons. We can easily build complex interfaces using the SDK; however, we need to first learn about *controllers* in order to create them.

Views are only one leg of the *Model-View-Controller* (MVC) programming paradigm adopted by the iOS SDK. A "programming paradigm" sounds intimidating, but MVC is actually fairly simple. The basic idea is that your code should have three types of objects: models to represent data, views to display those models, and controllers to process user input.

You can think of controllers as a layer between the user and the rest of your code. Their role is to interpret events and forward the changes to the relevant models and views. For example, tapping a button should be detected by a controller, which then increments a data property (model) and updates a label to reflect the change (view).

Controllers are instances of UIViewController in iOS. The SDK comes with several UIViewController subclasses with custom views and behavior to give every app the same look and feel. Controllers are absolutely central to iOS development, so we're going to take a look at how we use them.

3.1 Adding a New UIViewController

As the name suggests, UIViewControllers are objects that control a view. The UIViewController object stores the UIView it manages inside the view attribute. However, we generally don't use addSubview: to add this particular view to the screen; instead, various methods will often take the entire UIViewController object and adjust the view as necessary before adding it to a hierarchy.

Let's get started with a small project to see how UIViewController works, and you'll see what I mean. We'll create an app showing different ways to explore colors, specifically, motion create ColorViewer.

First we make the ./app/controllers directory (mkdir ./app/controllers). This is where we'll keep all of our controller classes. When building production-level apps, you should also add views and models subdirectories, but we won't be needing those right now.

Then we add a colors_controller.rb file in controllers. This will be our custom UIView-Controller subclass that's presented to the user. We'll start by setting its superclass and adding one short method.

controllers/ColorViewer/app/controllers/colors_controller.rb
```
class ColorsController < UIViewController
  def viewDidLoad
    super

    self.view.backgroundColor = UIColor.whiteColor

    @label = UILabel.alloc.initWithFrame(CGRectZero)
    @label.text = "Colors"
    @label.sizeToFit
    @label.center =
      [self.view.frame.size.width / 2,
       self.view.frame.size.height / 2]
    @label.autoresizingMask =
      UIViewAutoresizingFlexibleBottomMargin | UIViewAutoresizingFlexibleTopMargin
    self.view.addSubview(@label)
  end
end
```

Subclassing UIViewController always involves overriding viewDidLoad(). This method is called right after our controller's view has been created and is where we do whatever custom setup is necessary. For now, we just set the view's background color, add a label, and call it a day.

viewDidLoad() is one of the *view life-cycle* methods. Every controller's view goes through several stages: creation, appearance, disappearance, and destruction. You can add custom behaviors at each point using the corresponding life-cycle methods, but the most common is viewDidLoad().

Now that we have our controller, open our AppDelegate. We're going to create a UIWindow just like we did in the previous chapter, except we're now going to use the rootViewController=() method instead of addSubview:.

controllers/ColorViewer/app/app_delegate.rb

```ruby
class AppDelegate
  def application(application, didFinishLaunchingWithOptions:launchOptions)
    @window = UIWindow.alloc.initWithFrame(UIScreen.mainScreen.bounds)
    @window.makeKeyAndVisible

    @window.rootViewController =
      ColorsController.alloc.initWithNibName(nil, bundle: nil)

    true
  end
end
```

rootViewController=() will take the UIViewController and adjust the view's frame to fit the window. This lets us write our controller without hard-coding its size, making our controller reusable to other containers. As we said earlier, methods in which you pass a UIViewController are very common, as we'll soon see.

We instantiate UIViewControllers with initWithNibName:bundle:. This method can be used to load a controller from a .NIB/.XIB file created using Xcode's Interface Builder, but in this case, we passed nil, meaning the controller will be created programatically.[1]

initWithNibName:bundle: is the *designated initializer* of UIViewController. Whenever you want to create a controller, you *must* call this method at some point, either directly or inside the definitions of your custom initializers (such as controller.initWithSome:property:).

Let's run our app and check it out:

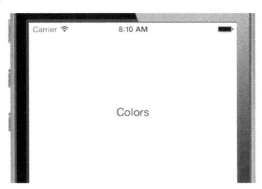

---

1. RubyMotion does support Interface Builder; simply add your .NIB or .XIB files to the project's ./resources directory. Using Interface Builder is beyond the scope of this text, but you can use the IB RubyGem (https://github.com/RubyMotion/ib) to help connect your Ruby- Motion code inside Interface Builder.

As we saw in Section 2.2, *Making Shapes and Colors*, on page 12, a few addSubview:s could have given us the same result, but using controllers creates a view that can easily fit into different containers. In fact, we're going to do just that with UINavigationController in Section 3.2, *Using Multiple Controllers with UINavigationController*, on page 30.

---

super and the Life-Cycle Methods

It's a good habit to call super() in viewDidLoad() and the other life-cycle methods. The default implementations of UIViewController can have important setup details, and you may experience some unexpected and hard-to-debug behavior if you don't call them; this is particularly true if you start subclassing UINavigationController or UITableViewController.

In some cases, notably viewDidAppear, Apple explicitly says you need to call the superclass method in any UIViewController subclass. Refer to Apple's documentation[a] for more details.

---

a. http://developer.apple.com/library/ios/#documentation/uikit/reference/UIViewController_Class/Reference/Reference.html

---

3.2 Using Multiple Controllers with UINavigationController

Although some iOS apps are famous for their unique visuals, most apps share a common set of interface elements and interactions included with the SDK. Typical apps will have a persistent top bar (usually blue) with a title and some buttons; these apps use an instance of UINavigationController, one of the standard *container controllers* in iOS.

Containers are UIViewController subclasses that manage many other *child* UIViewControllers. Kind of wild, right? Containers have a view just like normal controllers, to which their children controllers' views are added as subviews. Containers add their own UI around their children and resize their subviews accordingly. UINavigationController adds a navigation bar and fits the children controllers below, like this:

UINavigationController manages its children in a stack, pushing and popping views on and off the screen. Visually, new views are pushed in from the right, while old views are popped to the left. For example, Mail.app uses this to dig down from an inbox to an individual message. UINavigationController also automatically handles adding the back button and title for you; all you need to worry about is pushing and popping the controller objects you're using. Check out the following to see how Settings.app uses navigation controllers.

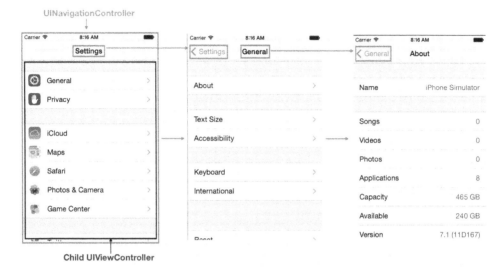

UINavigationController is pretty easy to integrate. In AppDelegate, we change our rootViewController() assignment to use a new UINavigationController.

controllers/ColorViewer_nav/app/app_delegate.rb

```
class AppDelegate
  def application(application, didFinishLaunchingWithOptions:launchOptions)
    @window = UIWindow.alloc.initWithFrame(UIScreen.mainScreen.bounds)
    @window.makeKeyAndVisible
    controller = ColorsController.alloc.initWithNibName(nil, bundle: nil)
    nav_controller =
      UINavigationController.alloc.initWithRootViewController(controller)
    @window.rootViewController = nav_controller
    true
  end
end
```

initWithRootViewController: will take the given controller and start the navigation stack with it. As we said earlier, the UINavigationController will handle adding and resizing this controller's view to fit to the appropriate size.

Before we run the app, we should make one more change in ColorsController. Every UIViewController has a title(), which UINavigationController uses to set the top bar's title.

controllers/ColorViewer_nav/app/controllers/colors_controller.rb

```
self.view.addSubview(@label)
self.title = "Colors"
```

Run and check out our slightly prettier app:

Excellent, let's make it do something. We'll add a few buttons to our view, each representing a color. When a button is tapped, we'll push a detail controller with its color as the background. Visually, a new controller will slide in from the right while the old ColorsController fades to the left.

First we need to add those buttons to the view at the end of viewDidLoad(). We're going to use some of Ruby's dynamic features to get this done, primarily the send() method. You can use any of the default UIColor helper methods like purpleColor() or yellowColor(), but we're going to stick with the basics.

controllers/ColorViewer_nav/app/controllers/colors_controller.rb
```ruby
["red", "green", "blue"].each_with_index do |color_text, index|
  color = UIColor.send("#{color_text}Color")
  button_width = 80

  button = UIButton.buttonWithType(UIButtonTypeSystem)
  button.setTitle(color_text, forState:UIControlStateNormal)
  button.setTitleColor(color, forState:UIControlStateNormal)
  button.sizeToFit
  button.frame = [
    [30 + index*(button_width + 10),
     @label.frame.origin.y + button.frame.size.height + 30],
    [80, button.frame.size.height]
  ]
  button.autoresizingMask =
    UIViewAutoresizingFlexibleBottomMargin | UIViewAutoresizingFlexibleTopMargin
  button.addTarget(self,
    action:"tap_#{color_text}",
    forControlEvents:UIControlEventTouchUpInside)
  self.view.addSubview(button)
end
```

See the color = UIColor.send("#{color_text}Color") trick? This lets us create the UIColor, button text, and button callback all with a single color_text variable. If you run our app now, you should see the three buttons just like this (but don't tap any quite yet):

Now on to implementing those button callbacks. For each of these, we'll need to push a new view controller onto our UINavigationController's stack. UIViewControllers happens to have a navigationController() property, which lets us access the parent UINavigationController. This navigationController() is automatically set whenever we add a view controller to a navigation stack, which we did with initWithRootView-Controller:. With that in mind, our button callbacks look something like this:

controllers/ColorViewer_nav/app/controllers/colors_controller.rb
```
def tap_red
  controller = ColorDetailController.alloc.initWithColor(UIColor.redColor)
  self.navigationController.pushViewController(controller, animated: true)
end
def tap_green
  controller = ColorDetailController.alloc.initWithColor(UIColor.greenColor)
  self.navigationController.pushViewController(controller, animated: true)
end
def tap_blue
  controller = ColorDetailController.alloc.initWithColor(UIColor.blueColor)
  self.navigationController.pushViewController(controller, animated: true)
end
```

We call pushViewController:animated: on the navigation controller, which pushes the passed controller onto the stack. By default, the navigation controller will also create a back button that will handle popping the frontmost child controller for us. If you need to do that programmatically, just call popViewControllerAnimated:() on UINavigationController.

We referenced a new ColorDetailController class, so let's implement that. First we create color_detail_controller.rb in ./app/controllers. We'll keep it simple and just set the title and background color.

controllers/ColorViewer_nav/app/controllers/color_detail_controller.rb
```
class ColorDetailController < UIViewController
  def initWithColor(color)
    self.initWithNibName(nil, bundle:nil)
    @color = color
    self.title = "Detail"
    self
  end
  def viewDidLoad
    super

    self.view.backgroundColor = @color
  end
end
```

Start by defining a new initializer, initWithColor:, which takes a UIColor as its argument. Our implementation of this method uses UIViewController's designated initializer initWithNibName:bundle:(), which is required for any controller initializer we write. You also need to return self from these functions, which should make sense given all the times we assign a variable from these methods (such as controller = UIViewController.alloc.initWithNibName(nil, bundle: nil)).

It's time to rake and play with our navigation stack. It should look like this:

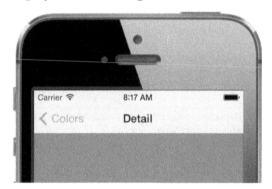

Check out the slick animations on the navigation bar, where the title simultaneously fades and slides as a new controller is pushed.

Many apps structure their interface using UINavigationController, where each pushed controller gradually reveals more detailed data. As you saw, there are only a couple of methods we need to implement that user interface. However, some apps need more than this kind of hierarchal layout. UITabBarController is another widely used container controller, and in Section 3.3, *Separating Controllers with UITabBarController*, on page 34 we're going to add it to our app.

3.3 Separating Controllers with UITabBarController

UITabBarController functions an awful lot like UINavigationController. The children controllers' views fit above the black tab bar, where each tab corresponds to one child. The Music app shows a tab with room for four controllers:

The fact that the More tab appears here indicates there are more than five children.

Unlike other containers, UITabBarControllers are *only* to be used as the rootViewController() of a UIWindow. You cannot push an instance of UITabBarController in pushViewController:animated:. From a user-experience perspective, this means you should use a tab bar only if contains very distinct and globally applicable controllers.

Just like UINavigationController, tab bars are easy to add. It just takes a small change to AppDelegate.

controllers/ColorViewer_tab/app/app_delegate.rb
```
controller = ColorsController.alloc.initWithNibName(nil, bundle: nil)
nav_controller =
  UINavigationController.alloc.initWithRootViewController(controller)
tab_controller =
  UITabBarController.alloc.initWithNibName(nil, bundle: nil)
tab_controller.viewControllers = [nav_controller]
@window.rootViewController = tab_controller
```

We create a UITabBarController like a normal UIViewController and set its viewControllers() to an array containing our navigation controller. The order of viewControllers() corresponds to the left-to-right order of the tabs.

When you run our app, you can see the top and bottom bars typical to most iOS apps implemented with a pretty small amount of code. Since it's not very helpful to have just one unstyled tab, let's fix that.

Every UIViewController has a tabBarItem() property, which accepts UITabBarItem, an object containing information about how to draw the view for the controller in the bottom tab bar. It is *not* a UIView but rather a plain object that the system uses to construct a view. We use the UITabBarItem to customize the icon, title, and other appearance options for the controller's tab.

The first step is to override initWithNibName:bundle: in ColorsController, and then we can create our UITabBarItem.

controllers/ColorViewer_tab/app/controllers/colors_controller.rb
```
def initWithNibName(name, bundle: bundle)
  super
  self.tabBarItem =
    UITabBarItem.alloc.initWithTitle(
      "Colors",
      image: nil,
      tag: 1)
  self
end
```

initWithTitle:image:tag: is one initializer for UITabBarItem, which we can use to set a custom image and title. tag: can be used to uniquely identify the tab bar item, but we won't be using it here. image should be a 30x30 black and transparent icon. Setting image to nil means we won't display images here.

You can also use the initWithTabBarSystemItem:tag: initializer to automatically set the title and image, assuming your tab corresponds to one of the default styles (such as Favorites or Contacts).

Why did we create our tab item in initWithNibName:bundle:? We want to create the tabBarItem() as soon as the controller exists, regardless of whether its view has been created yet. UITabBarController will load the views only when absolutely necessary, so if you wait to create the tab item in viewDidLoad(), then some controllers' items might not be set when the app is done launching.

One more thing! We should probably add another tab, right? We'll pretend this is the Top Color section, where we can view the most popular color. This way we can reuse our ColorDetailController.

controllers/ColorViewer_tab/app/app_delegate.rb
```
top_controller = ColorDetailController.alloc.initWithColor(UIColor.purpleColor)
top_controller.title = "Top Color"
top_nav_controller =
  UINavigationController.alloc.initWithRootViewController(top_controller)
tab_controller.viewControllers = [nav_controller, top_nav_controller]
```

Run the rake command once again, and *voila!* You should see a whole bunch of container controllers like this:

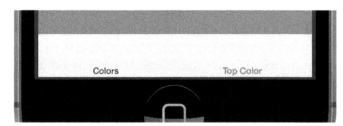

Before we add even more content to our app, let's take a moment to examine a few of the more subtle details of controllers.

The Edges of UIViewControllers

If you take a look at our second tab, you should notice that the purple background extends *underneath* the tab bar. In fact, all of our colored screens extend underneath the default iOS navigation elements. How does that happen?

Prior to iOS7, the navigation bar, tab bar, and other interface elements were opaque, so any interior controllers were sized to always be visible on the screen. But with the release of iOS7, many interface elements switched to a translucent visual effect, and Apple encourages apps to take advantage of that by layering their controllers and views.

This present a few problems: what if we don't want important items to be hidden below the bars? What if we don't want anything underneath? Apple provides a few methods on UIViewController to help us out.

If you don't want any part of your controller to show up underneath the navigation chrome, you can use the edgesForExtendedLayout() property of your UIViewController. For example, to prevent our second tab from leaking its color under the tab bar, we just change this property to UIRectEdgeNone:

controllers/ColorViewer_edge/app/app_delegate.rb
```
top_controller.title = "Top Color"
top_controller.edgesForExtendedLayout = UIRectEdgeNone
top_nav_controller =
  UINavigationController.alloc.initWithRootViewController(top_controller)
tab_controller.viewControllers = [nav_controller, top_nav_controller]
```

Try out the app and see how the purple is no longer slipping under the navigation bar or tab bar. We could even set the edgesForExtendedLayout() property somewhere in the definition of our UIViewController subclass, like viewDidLoad(), if we never wanted this class to extend its edges.

Our controllers don't do a whole lot, but you can see how these two classes form the building blocks of many iOS apps. UINavigationController and UITabBarController provide easy ways to organize many different parts of your app, but what if we really need to focus the user's attention on just one screen? Well, it turns out that we can also present controllers *modally* in front of all other controllers.

3.4 Presenting Modal UIViewControllers

Sometimes we want one controller to take up the entire screen to get a user's attention. For example, Mail.app's New Message screen appears on top of the usual inbox list, forcing the user to either complete the message or explicitly end the task. To accomplish this, we present the controller modally.

UIViewControllers allows us to present modal view controllers at any point in their life cycle. The key method is presentViewController:animated:completion:, which functions similarly to UINavigationController's pushViewController:animated:. The given controller will be presented above all other controllers in the app and will remain there until we invoke dismissViewControllerAnimated:completion:.

Let's present a modal controller from our Top Color controller. The presented controller will allow us to change the top color, which is definitely a task best done while the rest of the interface is obscured.

First we need a button in the navigation bar at the top of the screen that presents this modal controller. We saw UIButton in Chapter 2, *Filling the Screen with Views*, on page 11, but we need to use a different class for bar buttons: UIBarButtonItem. This isn't a subclass of UIView; instead, it's a plain-old Ruby object that we use to specify the text, image, and style of the bar button. The system will then take care of how to draw and add the button specifications as a view, much like UITabBarItem.

Let's add our bar button to our app. In ColorDetailController, we create the button item in viewDidLoad() like so:

```
controllers/ColorViewer_modal/app/controllers/color_detail_controller.rb
  rightButton =
    UIBarButtonItem.alloc.initWithTitle("Change",
      style: UIBarButtonItemStyleBordered,
      target:self,
      action:'change_color')
  self.navigationItem.rightBarButtonItem = rightButton
end
```

We create our UIBarButtonItem instance with a title and a style. The style property determines how our button looks: it can be plain, bordered, or "done" (play around to see the difference). We then set the new UIBarButtonItem as our controller's navigationItem()'s rightBarButtonItem(). Every UIViewController has a navigationItem(), which is how we access all the information displayed in the top bar. Again, note that UINavigationItem is *not* a UIView, so you cannot add new subviews to it.

We also assign a target and action in the initializer, which function in the same manner as when we call addTarget:action:forControlEvents: on a UIButton.

We haven't implemented change_color() yet, so let's get to it. To make modal controllers stand out even more, it's a common practice to wrap them in a small UINavigationController. All we need to do is call presentViewController:animated:completion: with that controller, so it's short and sweet.

```
controllers/ColorViewer_modal/app/controllers/color_detail_controller.rb
def change_color
  controller = ChangeColorController.alloc.initWithNibName(nil, bundle:nil)
  controller.color_detail_controller = self
  self.presentViewController(
    UINavigationController.alloc.initWithRootViewController(controller),
    animated:true,
    completion: lambda {})
end
```

Everything looks normal except the lambda in completion. Just like the view animations we saw in Section 2.4, *Animating Views*, on page 16, presenting controllers take an anonymous callback function. We don't need to do any special behavior right now, but it's there if you ever need it. Our presented controller is a new ChangeColorController object, which is a class we don't have yet.

Create change_color_controller.rb in ./app/controllers; this will be the controller that we actually present. It won't be a super-complicated class: we'll add a text field for the user to enter the color, as well as a button to enact that change. But before all that, we need to set up the plumbing.

controllers/ColorViewer_modal/app/controllers/change_color_controller.rb
```
class ChangeColorController < UIViewController
  attr_accessor :color_detail_controller
```

We start with Ruby's nifty attr_accessor() to create the methods color_detail_controller() and color_detail_controller=(). We need these so we can easily store a reference to the ColorDetailController whose color we're changing.

The ChangeColorController modal view also needs a UITextField and UIButton, both of which we covered in Chapter 2, *Filling the Screen with Views*, on page 11. When the button is tapped, we'll take whatever is in the text field and use that to create a UIColor that ColorDetailController can use. Just like the other controllers, this logic belongs in viewDidLoad().

controllers/ColorViewer_modal/app/controllers/change_color_controller.rb
```
def viewDidLoad
  super
  self.title = "Change Color"
  self.view.backgroundColor = UIColor.whiteColor
  @text_field = UITextField.alloc.initWithFrame(CGRectZero)
  @text_field.borderStyle = UITextBorderStyleRoundedRect
  @text_field.textAlignment = UITextAlignmentCenter
  @text_field.placeholder = "Enter a color"
  @text_field.frame = [CGPointZero, [150,32]]
  @text_field.center =
    [self.view.frame.size.width / 2, self.view.frame.size.height / 2 - 170]
  self.view.addSubview(@text_field)
  @button = UIButton.buttonWithType(UIButtonTypeSystem)
  @button.setTitle("Change", forState:UIControlStateNormal)
  @button.frame = [[
    @text_field.frame.origin.x,
    @text_field.frame.origin.y + @text_field.frame.size.height + 10
    ],
    @text_field.frame.size]
  self.view.addSubview(@button)
  @button.addTarget(self,
    action:"change_color",
    forControlEvents:UIControlEventTouchUpInside)
end
```

It's lengthy, but most of the code just lays out our views. We position @text_field slightly above the center of the view and then position @button right below. Finally, we set our callback to be change_color(), so we need to write that too.

controllers/ColorViewer_modal/app/controllers/change_color_controller.rb
```ruby
  def change_color
    color_text = @text_field.text
    color_text ||= ""
    color_text = color_text.downcase
    color_method = "#{color_text}Color"
    if UIColor.respond_to?(color_method)
      color = UIColor.send(color_method)
      self.color_detail_controller.view.backgroundColor = color
      self.dismissViewControllerAnimated(true, completion: nil)
      return
    end

    @text_field.text = "Error!"
  end
end
```

First we try to generate a UIColor from @text_field.text. We use respond_to?() to catch invalid colors (like "catdogColor"), but if nothing bad happens, then we forge ahead. We grab a reference to our ColorDetailController using color_detail_controller(), set its background color, and then dismiss ourselves with dismissViewControllerAnimated:completion:(). Not bad at all, right?

Take our app for a spin, and everything should go smoothly, as in the following figure:

Figure 6—Changing the color from a text field

Make sure to test our exception handling in ChangeColorController, as well some less-known UIColor helpers like magenta or cyan.

We made some really tangible progress in this chapter as we built software that looks and acts like what's expected on iOS. The code we went through should give you a better idea about the biggest difference between the iOS APIs and plain Ruby: the original Objective-C method names are very verbose. Need I say even more than UIViewAutoresizingFlexibleBottomMargin? Thankfully, we pulled a few Ruby tricks with UIColor.send and saved even more boilerplate by removing the need for header files and complex class definitions just with attr_accessor().

Our apps can now be organized using the standard UI patterns, but that last example showed a significant gap in our knowledge: changing data can be messy. Passing the ColorDetailController as a property and altering its view directly is less than desirable. We're going to cover a more automatic way of handling those kind of data changes and more in Chapter 4, *Representing Data with Models*, on page 43.

Representing Data with Models

We've covered views and controllers, but where's the love for the *M* in MVC? Well, wait no more, because we're going to dive into models. We now understand views and controllers, but in practice models will play just as big of a role as the sexier, user-facing code.

In iOS, there are two big components to models: *CoreData* and, well, everything else. CoreData is an iOS object graph and persistence framework, sort of similar to ActiveRecord in Rails-land. It's an incredibly powerful framework to save and query objects using a database, so it deserves a chapter or even a book onto its own. But even without touching CoreData, we can do a whole lot with just "everything else" about models. It's time to get down to business.

4.1 Writing Basic Models

Unlike controllers and views, there's no default superclass that models inherit from; they're just plain-old Ruby objects. We can use the standard attr_accessor, reader, and writer functions to declare getter and setter methods, which sometimes are all you need.

Many apps have users and profiles, so let's work through a portion of that sort of app and use some nice models. Let's create a new project (such as motion create UserProfile) and two subdirectories within ./app: models and controllers. We're not only going to cover models; we're going to keep building on what we already know.

Our app will let us create, view, and edit users. Since we're doing an awful lot of work with users, this sounds like a good place to begin with our first model. To get started, we first create user.rb in ./app/models. For now, each user will have a name, email, and ID. This is a pretty basic implementation of User:

```
models/UserProfile/app/models/user.rb
class User
  attr_accessor :id
  attr_accessor :name
  attr_accessor :email
end
```

That's all we need for now. Let's add our first controller, which we can use to view a user. Create user_controller.rb in app/controllers, which will be a subclass of UIViewController. We'll use a custom initializer that takes a User and fills the UI appropriately. Sound good? Let's start with that initializer.

```
models/UserProfile/app/controllers/user_controller.rb
class UserController < UIViewController
  attr_accessor :user

  def initWithUser(user)
    initWithNibName(nil, bundle:nil)
    self.user = user
    self.edgesForExtendedLayout = UIRectEdgeNone
    self
  end
```

Pretty typical initializer, isn't it? We're going to assume the user passed is a User object, but we'll worry about that later. Next, we need to set up the view. This will be kind of lengthy, but that's nothing new for us at this point.

For each of our User properties, we'll create two labels: one to tell us what value we're looking at ("Email") and one right beside it that presents the value of that property for our user ("clay@mail.com"). Since we are using Ruby's nifty send(), we can do this in one loop.

```
models/UserProfile/app/controllers/user_controller.rb
  def viewDidLoad
    super

    self.view.backgroundColor = UIColor.whiteColor

    last_label = nil
    ["id", "name", "email"].each do |prop|
      label = UILabel.alloc.initWithFrame(CGRectZero)
      label.text = "#{prop.capitalize}:"

      label.sizeToFit
      if last_label
        label.frame = [
          [last_label.frame.origin.x,
            last_label.frame.origin.y + last_label.frame.size.height],
          label.frame.size]
```

```
    else
      label.frame = [[10, 10], label.frame.size]
    end
    last_label = label

    self.view.addSubview(label)

    value = UILabel.alloc.initWithFrame(CGRectZero)
    value.text = self.user.send(prop)
    value.sizeToFit
    value.frame = [
      [label.frame.origin.x + label.frame.size.width + 10, label.frame.origin.y],
      value.frame.size]
    self.view.addSubview(value)
  end
  self.title = self.user.name
 end
end
```

Don't get overwhelmed! We just created two UILabels for each property and laid them out nicely.

Before we run our app, we need to set up our AppDelegate to get the controller on the screen. First, we create a new User in app_delegate.rb and use it to initialize a UserController. We're going to wrap it in a UINavigationController so we get the nice effect with self.title.

models/UserProfile/app/app_delegate.rb
```
class AppDelegate
  def application(application, didFinishLaunchingWithOptions:launchOptions)
    @window = UIWindow.alloc.initWithFrame(UIScreen.mainScreen.bounds)

    @user = User.new
    @user.id = "123"
    @user.name = "Clay"
    @user.email = "clay@mail.com"
    @user_controller = UserController.alloc.initWithUser(@user)
    @nav_controller =
        UINavigationController.alloc.initWithRootViewController(@user_controller)
    @window.rootViewController = @nav_controller
    @window.makeKeyAndVisible
    true
  end
end
```

Let's give it a rake and see what happens! Your app should look something like Figure 7, *A controller for our User model*, on page 46. Our view isn't the prettiest, but a good designer could dress up even this limited information into something shippable.

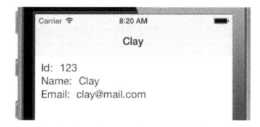

Figure 7—A controller for our User model

This is great, but this User class didn't help out all that much. We're going to give it some abilities that will make our code much more flexible while showing off what smart models can do.

4.2 Preparing Scalable Models

The word *scalable* gets thrown around a lot, usually in terms of databases or web back ends, but client-side code can easily become unscalable as well. For instance, if we wanted to add more attributes to our User, we'd have to make changes in three places: our class definition, our controller, and where we instantiate our objects. In a world where engineers move fast, having all this overhead can be a big time sink.

I wouldn't bring this up if there weren't a better way. We're going to use a nice Ruby trick that lets our models become more flexible and readies them for a typical API. In user.rb, change our three attr_acessor() lines into this:

```
models/UserProfile_flex/app/models/user.rb
class User
  PROPERTIES = [:id, :name, :email]
  attr_accessor *PROPERTIES
```

Nifty, right? Now we have one data structure containing our desired properties, instead of multiple independent lines. This lets us refactor code that should apply to all properties into more loops like PROPERTIES.each. For example, we can now make our User initializable with a hash.

```
models/UserProfile_flex/app/models/user.rb
def initialize(properties = {})
  properties.each do |key, value|
    if PROPERTIES.member? key.to_sym
      self.send("#{key}=", value)
    end
  end
end
```

This lets us initialize users in one line instead of needing one line for every property. Plus, when we add new properties, the initializer code still works. Let's start updating our old code to use this concept. In UserController, change our properties from being hard-coded to using User::PROPERTIES.

```
models/UserProfile_flex/app/controllers/user_controller.rb
last_label = nil
User::PROPERTIES.each do |prop|
  label = UILabel.alloc.initWithFrame(CGRectZero)
```

While we're at it, let's change how we created the @user instance variable in AppDelegate.

```
models/UserProfile_flex/app/app_delegate.rb
@window = UIWindow.alloc.initWithFrame(UIScreen.mainScreen.bounds)

@user = User.new(id: "123", name: "Clay", email: "clay@mail.com")

@user_controller = UserController.alloc.initWithUser(@user)
```

If you run our app now, nothing *looks* different, but under the hood we've made some important changes. Just to show how useful this is, let's add a new phone property to User so that we can see how we just change the input when creating a new object.

```
models/UserProfile_flex2/app/models/user.rb
class User
  PROPERTIES = [:id, :name, :email, :phone]
  attr_accessor *PROPERTIES
```

```
models/UserProfile_flex2/app/app_delegate.rb
@window = UIWindow.alloc.initWithFrame(UIScreen.mainScreen.bounds)

@user = User.new(id: "123", name: "Clay",
    email: "clay@mail.com", phone: "555-555-5555")
```

Once you run rake, you'll see that we now have a matching UI element for our new property without having to add new code in the controller. This is a great example of how smart(er) models can save us time, but what if we wanted to edit this data?

4.3 Changing Models with Key-Value Observing

Making our model code flexible is the low-hanging fruit, and keeping a view up-to-date with a model is usually a hard problem. A model object has to be aware of all the views accessing its data, which just leads to all sorts of problems and code spaghetti. Imagine a world where it all just works: you change the name on a User, and the appropriate label instantly reflects those changes...no code spaghetti or zero-ing references.

I'm talking this up so much because it is possible to implement. iOS has a concept of *key-value observing* (KVO). Built into the frameworks is a system by which one object can passively observe changes in properties of another object. Those properties are referred to by *keys*, which typically correspond to the variable name of the attribute.

In our User case, our controller would observe the "name" key of the User. In the callback for that observation, we would reset the label's text to reflect the new name value.

I don't know about you, but I'm anxious to write some code. But not so fast! The original implementation of KVO isn't very Ruby-like. Thankfully, there is a popular RubyGem called BubbleWrap (http://bubblewrap.io) that "wraps" many Objective-C APIs into idiomatic Ruby structures. To install it, run gem install bubble-wrap in your shell and add require "bubble-wrap" to your Rakefile.

models/UserProfile_kvo/Rakefile
```
$:.unshift("/Library/RubyMotion/lib")
require 'motion/project/template/ios'
➤ require 'bubble-wrap'

begin
  require 'bundler'
  Bundler.require
rescue LoadError
end
```

Note here that the order of these statements does matter: BubbleWrap's require should come before the block involving Bundler.

BubbleWrap is an extensive library with many different features, so definitely browse its documentation sometime. For now, we're just going to use the wrappers it creates for key-value observing on our model. We'll be working with UserController, so let's open that and include BubbleWrap's KVO module.

models/UserProfile_kvo/app/controllers/user_controller.rb
```
class UserController < UIViewController
➤   include BubbleWrap::KVO
  attr_accessor :user
```

This is a nice and easy first step. This gives us access to the observe(object, key) method, which we'll use right now. In the loop over User::PROPERTIES, we're going to observe each property and update the value label accordingly. After we initialize the label, add the observe() method.

Third-Party Libraries and RubyMotion

RubyMotion does not currently support require in our source code files, so we need to use the RubyGems package manager (http://rubygems.org/) to bundle external libraries. RubyGems comes pre-installed on most versions of OS X, but you should visit the RubyGems website to download the latest version if necessary.

Installing RubyMotion-specific gems isn't any different from normal desktop gems: in your terminal, run gem install [gem name]. It will be downloaded like any other gem and can be required in your project's Rakefile; however, there will usually be an exception if you try to use it in non-RubyMotion apps. Since these are normal RubyGems, you can use Bundler (http://bundler.io/) to manage your gems like any other Ruby project.

models/UserProfile_kvo/app/controllers/user_controller.rb
```ruby
value = UILabel.alloc.initWithFrame(CGRectZero)
value.text = self.user.send(prop)
observe(self.user, prop) do |old_value, new_value|
  value.text = new_value
  value.sizeToFit
end
```

```ruby
value.sizeToFit
```

We pass it the object we want to observe (self.user) and the key we want updates about (prop). The callback block returns both the old and new values for the property, which could be useful if we were making more discriminatory UI updates.

Since the controller's title is originally set to the user's name, it's a nice idea to update that as the name changes.

models/UserProfile_kvo/app/controllers/user_controller.rb
```ruby
self.title = self.user.name
  observe(self.user, "name") do |old_value, new_value|
    self.title = new_value
  end
```

Finally, we need to do a quick cleanup by overriding viewDidUnload(), one of the controller life-cycle methods. Since we're creating these observations in view-DidLoad(), we need to mirror their un-observing in the counterpart method.

models/UserProfile_kvo/app/controllers/user_controller.rb
```ruby
def viewDidUnload
  unobserve_all
  super
end
```

Whew, all done. Let's play with our app and actually see the fruits of our labor. Go ahead and rake and get ready to use the debugger!

In the interactive debugger, grab the @user instance variable of our AppDelegate. BubbleWrap includes a nifty shortcut for grabbing the app's delegate, which we can now use.

```
(main)> user = App.delegate.instance_variable_get("@user")
=> #<NSKVONotifying_User @id="123", @email="clay@mail.com", @phone="555-555-5555">
```

Kind of a weird class name, isn't it? I won't get too technical, but under the hood KVO does a lot of tricks involving dynamically subclassing your observed objects. But it is proof that our user is being observed! So, let's make some changes:

```
(main)> user.email = "my_new_email@host.com"
=> "my_new_email@host.com"
(main)> user.name = "Charlie"
=> "Charlie"
```

In *Creating a New App*, we learned to make changes to the UI using the debugger, but now we're not even playing with the view objects! Everything just works. This is an incredibly powerful asset in your toolbox when it comes to making more complex, reactive apps.

So, now that we can make all these changes and synchronize the UI, how can we save them? If we quit the app right now and restart, our old "Clay" user will still be hanging around. And I personally have no problem with that, but most apps will want to save the user's changes.

4.4 Saving Data with NSUserDefaults and NSCoding

Applications generally have long-lasting consequences: we take a picture, create a presentation, or just unlock a new level. iOS will try to keep your app in memory for a reasonable amount of time, but eventually you need to permanently save something to the disk. There are several ways of doing this, ranging from writing files to using a SQLite database. We're going to use something in the middle: NSUserDefaults.

NSUserDefaults lets us persist basic objects (strings, numbers, arrays, and hashes) through a simple key-value interface. It handles the serialization mechanics for us, saving you the trouble of constructing a custom file serialization scheme. But it has one more trick up its sleeve: coupled with NSCoding, we can actually save arbitrary objects, not just the primitives classes.

An NSCoding-compliant object implements two specific methods that describe how to save and restore its properties into primitive objects. We can then take this collection of primitive objects and turn it into raw data, which NSUserDefaults understands. That all sounds a bit heady, so let's try implementing this fancy stuff in our app to get a better idea of what it can do.

First let's make our User NSCoding compliant. We need to add two new methods: initWithCoder(decoder) and encodeWithCoder(encoder). When we need to serialize and deserialize our object, these methods will be called. The objects they pass as arguments have simple APIs for retrieving and setting values. For our User, they look something like this:

models/UserProfile_persist/app/models/user.rb
```ruby
def initWithCoder(decoder)
  self.init
  PROPERTIES.each do |prop|
    saved_value = decoder.decodeObjectForKey(prop.to_s)
    self.send("#{prop}=", saved_value)
  end
  self
end
def encodeWithCoder(encoder)
  PROPERTIES.each do |prop|
    encoder.encodeObject(self.send(prop), forKey: prop.to_s)
  end
end
```

initWithCoder(decoder) is called when we want to deserialize our object out of the decoder instance. Thus, we use decodeObjectForKey(key) on all of our PROPERTIES (see how that keeps making our life easier!).

Conversely, encodeWithCoder(encoder) gets called when we want to save our object and encode its properties with encodeObject:forKey:. The values and keys we use here are exactly those we use in initWithCoder:.

Our User can be encoded and decoded, but now what? We want to save and load the AppDelegate's @user, which requires us to use the NSUserDefaults. If we're using primitive types like strings or arrays, saving them directly just works:

```ruby
defaults = NSUserDefaults.standardUserDefaults
defaults["some_array"] = [1,2,3]
defaults["some_number"] = 4

some_name = defaults["some_name"]
```

However, putting NSCoding objects such as @user into NSUserDefaults requires NSKeyedArchiver and NSKeyedUnarchiver. These two classes take NSCoding objects and transform them into instances of NSData, which can be safely stored or retrieved from the defaults like normal. NSKeyedArchiver uses the encodeWithCoder:

we implemented earlier, while NSKeyedUnarchiver uses initWithCoder:. That's a lot of NS prefixes, I know. Here's what an NSCoding serialization looks like:

```
my_object = # some NSCoding-compliant object
defaults = NSUserDefaults.standardUserDefaults
defaults["some_object"] = NSKeyedArchiver.archivedDataWithRootObject(my_object)

my_saved_data = defaults["some_object"]
my_saved_object = NSKeyedUnarchiver.unarchiveObjectWithData(my_saved_data)
```

That's going to get really tedious really fast to repeat everywhere for all our Users, so let's wrap it in some helper methods.

```
models/UserProfile_persist/app/models/user.rb
USER_KEY = "user"
def save
  defaults = NSUserDefaults.standardUserDefaults
  defaults[USER_KEY] = NSKeyedArchiver.archivedDataWithRootObject(self)
end
def self.load
  defaults = NSUserDefaults.standardUserDefaults
  data = defaults[USER_KEY]
  # protect against nil case
  NSKeyedUnarchiver.unarchiveObjectWithData(data) if data
end
```

Ah, much better. In a production app, our USER_KEY would probably be a function of the user's id, but since we have only one user in our app, it's not a big deal. All we have to do now is save and load our object when the app opens and closes.

```
models/UserProfile_persist/app/app_delegate.rb
@window = UIWindow.alloc.initWithFrame(UIScreen.mainScreen.bounds)
@user = User.load
@user ||= User.new(id: "123", name: "Clay",
    email: "clay@mail.com", phone: "555-555-5555")
@user_controller = UserController.alloc.initWithUser(@user)
```

```
models/UserProfile_persist/app/app_delegate.rb
def applicationDidEnterBackground(application)
  @user.save
end
```

In addition to application:didFinishLaunchingWithOptions:, the application delegate can respond to many more application life-cycle methods, similar to UIViewController. Apple recommends we save user data after the application has entered the background (as not to accidentally freeze the interface), so we use that callback to call @user.save.

Why don't you take our app for a spin, alter some @user properties, and then close our app? In fact, hold down the home button on the simulator and force-quit it with the red icon just to make sure we really got it. Open it back up, and your changes should reappear!

This small app had only one controller, but you can see how KVO and NSUserDefaults scales to more complex objects and relationships. But what about our user interfaces? How can we display hundreds or thousands of models on the screen and keep our app running smoothly? Well, that's where the incredibly versatile UITableView comes into play in *Showing Data with Table Views*.

CHAPTER 5

Showing Data with Table Views

We've gone through the basics of models, views, and controllers in iOS development, but now we're going to dig a little deeper. The UITableView class is so central to most iOS apps that it deserves a whole book to itself, but a chapter will have to suffice for now.[1]

Although they look very different, the Mail and SMS apps both use UITableViews to display their content:

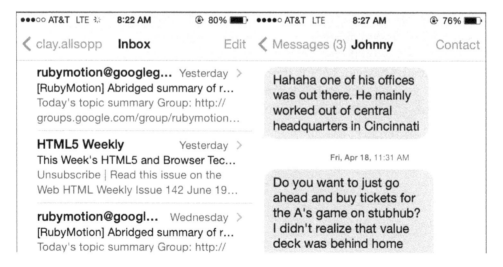

When you look through the default apps that come on your phone or iPad, you'll see something that's even more universal than blue and black tab bars: scrolling lists. And boy, do they scroll, right? The Phone.app smoothly scrolls through an enormous list of contacts; SMS.app practically flies through all your

---

1. Read the UITableView reference for detailed information: http://developer.apple.com/library/ios/
    #documentation/uikit/reference/UITableView_Class/Reference/Reference.html.

blue and green bubbles that are messages. Visually, they are quite distinct, but they're built with the same tool: UITableView.

UITableView manages content that can scroll far beyond the bounds of the screen. It's designed to present rows upon rows of similar-looking views, each populated with slightly different information. There is a simple reason why these lists can scale from nine to nine hundred items: table views use only a fixed number of subviews in memory at any given time. That sounds complicated, so I'll break it down.

In your Mail.app, you could have hundreds of emails in your inbox, but only a portion of them are on the screen at any given time (about six). When you scroll, the topmost emails eventually move off the screen. The table view will actually take these hidden views and move them to the *opposite* end of the screen. So, those message views that move off the top of the screen get repositioned below the screen and have their content replaced with the appropriate data for that new position. The table view can move its rows around incredibly fast, so even if there are hundreds of items in your table, only a handful of row views are kept in memory.

So, how do we take advantage of this great class? UITableView is a view like any other UIView, so we have to add it as a subview with the appropriate frame. We also need to assign it delegate and dataSource objects. These objects must implement a few methods, with the option of implementing more for finetuned control. They can be any kind of object but are commonly controllers. The table view will then call those methods to get the information it needs to form the rows it needs. Doesn't sound bad, right? Let's get to the code.

5.1 Starting with Simple Rows

For now, our task is simple: display a table with all the letters of the alphabet:

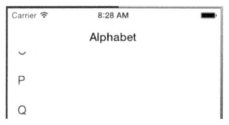

Last I checked there are twenty-six of them, so that's enough to show off the fancy row reuse that happens with table views. The table is going to be part of our app's lone controller, so there isn't a lot of setup to worry about this time.

First we create a new project called TableFun, and the controllers folder in ./app. Our controller will be AlphabetController, so let's store it in ./app/controllers/alphabet_controller.rb. Our initial implementation should look like this:

tables/TableFun/app/controllers/alphabet_controller.rb
```
class AlphabetController < UIViewController
  def viewDidLoad
    super
    self.title = "Alphabet"
    @table = UITableView.alloc.initWithFrame(self.view.bounds)
    @table.autoresizingMask = UIViewAutoresizingFlexibleHeight
    self.view.addSubview(@table)
  end
end
```

Looks like a typical controller so far, doesn't it? We store our table view in the @table instance variable and add it like a normal subview, sized to fit the controller. The one new thing here is autoresizingMask(), which is a bitmask of UIViewAutoResizing constants. These are used to define how a view changes when its parent view is being resized; in this case, we allow the height of the table to be flexible.

Next, we need to set up a window and add the AlphabetController as its root controller. Let's mosey on over to AppDelegate and add our simple setup.

tables/TableFun/app/app_delegate.rb
```
class AppDelegate
  def application(application, didFinishLaunchingWithOptions:launchOptions)
    @window = UIWindow.alloc.initWithFrame(UIScreen.mainScreen.bounds)
    @alphabet_controller = AlphabetController.alloc.initWithNibName(nil, bundle:nil)
    @window.rootViewController =
      UINavigationController.alloc.initWithRootViewController(@alphabet_controller)
    @window.makeKeyAndVisible
    true
  end
end
```

This actually adds it as the root of a UINavigationController, which gives it a nice UI. If you rake now, you'll see a lovely but empty table view. Let's fill that with some data.

First we need to set our AlphabetController as the table's dataSource at the end of viewDidLoad.

tables/TableFun_basic/app/controllers/alphabet_controller.rb
```
    self.view.addSubview(@table)

➤   @table.dataSource = self
```

A table's dataSource *must* implement the following methods, or else the app will throw an exception:

- tableView:cellForRowAtIndexPath:
- tableView:numberOfRowsInSection:

numberOfRowsInSection: should return exactly what it sounds like. cellForRowAtIndex-Path:'s job is to either create a new cell or recycle an offscreen one.

cellForRowAtIndexPath: uses two unfamiliar objects: indexPath is an instance of NSIndexPath, and the method is supposed to return a UITableViewCell.

There are actually a slew of methods the data source can implement, but the simplest needs only these. They should be defined in your controller and should follow this form:

```
def tableView(tableView, numberOfRowsInSection: section)
  # return the number of rows
end
def tableView(tableView, cellForRowAtIndexPath: indexPath)
  # return the UITableViewCell for the row
end
```

NSIndexPaths have section and row properties, which we use to describe the location of the row we're setting up. UITableViewCell is the UIView subclass that actually gets displayed in our table. Its default implementation comes with some useful properties, including a textLabel and an imageView. You can subclass UITableViewCell to add a custom look and behaviors, but we can get something useful using those default subview properties.

Here's a more complete implementation with the boilerplate filled in:

```
def tableView(tableView, cellForRowAtIndexPath: indexPath)
  @reuseIdentifier ||= "CELL_IDENTIFIER"

  cell = tableView.dequeueReusableCellWithIdentifier(@reuseIdentifier)
  cell ||= UITableViewCell.alloc.initWithStyle(
      UITableViewCellStyleDefault,
      reuseIdentifier:@reuseIdentifier)

  # put your data in the cell

  cell
end
```

Wait, what's this reuseIdentifier business? Well, UITableView does its reuse magic by giving an "identifier" to each type of cell used in the table. When you grab a cell for reuse, you actually grab it from a pool of cells with the desired identifier. If you have two visually distinct types of rows, you should have two identifiers. If for some reason you had a unique identifier for every row, none of the cells could get reused because there would only ever be one cell in each pool. In our app, all of the cells will look alike, so we need only one global identifier.

We use dequeueReusableCellWithIdentifier: to grab a UITableViewCell from the identifier's pool; if we don't have one, then the ||= will create a new UITableViewCell with a given style and the reuseIdentifier for the cell. The style determines which properties the cell has, such as subtitles and secondary values; for now, we're going to stick with the default.

||= who?

You might not be familiar with the ||= operator. In Ruby, a ||= b means "If a is nil or false, then assign it the value of b." It's a good way to give a default value to an object or ensure that it's assigned only once, as in the previous example.

Let's put all this together in our AlphabetController with some actual data. In viewDidLoad(), let's initialize an array to use as the titles of our rows. Since Ruby is awesome, we can use a quick one-liner.

tables/TableFun_basic/app/controllers/alphabet_controller.rb
```
  @table.dataSource = self

➤  @data = ("A".."Z").to_a
end
```

This will create an array containing all the uppercase letters of the alphabet. Nifty, huh? Now we need to fill in our two data source methods.

numberOfRowsInSection: is pretty straightforward, since we are working with pretty simple @data.

tables/TableFun_basic/app/controllers/alphabet_controller.rb
```
def tableView(tableView, numberOfRowsInSection: section)
  @data.count
end
```

We said we're dealing with table rows, so you may be wondering about the sections. A table view is actually made out of many sections, each filled with many rows. These sections are normally separated by the semitransparent gray bars stuck to the top of a table. Right now we have only one section, so no other logic is necessary.

In cellForRowAtIndexPath:, we simply set our cell's textLabel (an instance of UILabel) to use the corresponding string from @data.

```
tables/TableFun_basic/app/controllers/alphabet_controller.rb
def tableView(tableView, cellForRowAtIndexPath: indexPath)
  @reuseIdentifier ||= "CELL_IDENTIFIER"

  cell = tableView.dequeueReusableCellWithIdentifier(@reuseIdentifier)
  cell ||= UITableViewCell.alloc.initWithStyle(
      UITableViewCellStyleDefault,
      reuseIdentifier:@reuseIdentifier)

  cell.textLabel.text = @data[indexPath.row]
  cell
end
```

Now let's take our app for a spin. If you weren't familiar with the twenty-six letters of the alphabet, then now would be a great time to learn (well, except that our table just statically displays this data and isn't very interactive). I guess we should fix that, right?

5.2 Interacting with UITableViews

Currently our AlphabetController is the dataSource of the table, but remember how tables also have a delegate? We need to make our controller the delegate if we want it to respond when someone taps the row.

The table view's dataSource methods are generally about supplying information to a table; the delegate methods concern themselves with what happens after the table is set up and how the user interacts with it. We'll use one of these delegate methods to push a new view controller when a row is tapped.

We start by making our controller the delegate in viewDidLoad(), just like dataSource.

```
tables/TableFun_interact/app/controllers/alphabet_controller.rb
@table.dataSource = self
@table.delegate = self
```

There aren't any required methods of a table's delegate; you're free to implement any (or none) depending on what you need. We're going to use the tableView:did-SelectRowAtIndexPath: method to figure out when a row was tapped by the user, like so:

```
tables/TableFun_interact/app/controllers/alphabet_controller.rb
def tableView(tableView, didSelectRowAtIndexPath:indexPath)
  tableView.deselectRowAtIndexPath(indexPath, animated: true)
  letter = @data[indexPath.row]

  controller = UIViewController.alloc.initWithNibName(nil, bundle:nil)
  controller.view.backgroundColor = UIColor.whiteColor
  controller.title = letter
  label = UILabel.alloc.initWithFrame(CGRectZero)
```

```
    label.text = letter
    label.sizeToFit
    label.center = [controller.view.frame.size.width / 2,
      controller.view.frame.size.height / 2]
    controller.view.addSubview(label)
    self.navigationController.pushViewController(controller, animated:true)
  end
```

We need to use deselectRowAtIndexPath:animated: because by default UITableView will keep a row highlighted in blue once the user taps it. After that, we construct a basic UIViewController with a label for our letter and push it onto the stack.

At this point, our example works, but it's missing one subtle interface element: the right-pointing gray chevron on our table rows. These arrows are used to indicate that tapping a row will push a new controller. These icons and any other UIView are referred to as the table cell's *accessory*. iOS includes some accessories like the gray chevron, but we can assign arbitrary views for each cell.

Let's head back to tableView:cellForRowAtIndexPath: to assign each cell the UITableViewCellAccessoryDisclosureIndicator accessory. That *is* a long variable name, isn't it?

tables/TableFun_interact/app/controllers/alphabet_controller.rb

```
    cell = tableView.dequeueReusableCellWithIdentifier(@reuseIdentifier)
➤   cell ||= UITableViewCell.alloc.initWithStyle(
➤       UITableViewCellStyleDefault,
➤       reuseIdentifier:@reuseIdentifier)
➤   cell.accessoryType = UITableViewCellAccessoryDisclosureIndicator
```

Great! Let's give our app a whirl and move around! Our table should look like this:

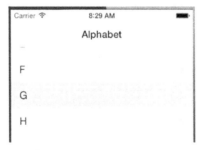

We haven't written a ton of code, but already it is a very familiar (and fast) interface. Much like Mail.app, we can tap table rows and dig deeper into a detail controller. But there's another common interface element missing from this example: sections.

5.3 Sections and Indexing Tables

If you look at Phone.app or Contacts.app, you'll notice that the alphabetical listing of your contacts has sticky headers at the top of the screen that indicate which part of the list you're viewing. These are the default UITableView section dividers. You'll also notice a bar on the right side that acts as a quick index for these sections, which is another built-in element of sections in UITableView. We're going to learn about both and add them to our app.

Adding sections to a table is pretty easy: implement a few more methods and add some additional logic in the existing dataSource implementations. Ideally, your data should be structured as a hash where the keys are your section titles and the values are rows contained in that section. Take our alphabet as an example.

```
@data = {
  "A" => ["Adam", "Apple"],
  "B" => ["Barry", "Berry"],
  "C" => ["Carlos", "Charlie"],
  ...
}
```

You get the idea. Your data doesn't need to be a hash, but you should have a way of referencing sections by their numerical index and then accessing the row data inside each section using a secondary index.

It's often useful to just create some helper methods in your dataSource so you don't repeat a lot of the same lookup code, which can be a pain if you change your structure later. In the AlphabetController, add the following:

```
tables/TableFun_sections/app/controllers/alphabet_controller.rb
def sections
  @data.keys.sort
end

def rows_for_section(section_index)
  @data[self.sections[section_index]]
end

def row_for_index_path(index_path)
  rows_for_section(index_path.section)[index_path.row]
end
```

These just abstract NSIndexPath-based access to our data. Again, it's not absolutely necessary, but if we ever change our data structure, then we would need to change only these three methods.

Now we need to implement a new dataSource method, numberOfSectionsInTableView:, which returns exactly what it says. Using the previous convenience methods, we implement it like so:

tables/TableFun_sections/app/controllers/alphabet_controller.rb
```ruby
def numberOfSectionsInTableView(tableView)
  self.sections.count
end
```

Now we need to go back and update our original dataSource and delegate methods to use the new data structure. These are all pretty easy changes thanks to our helper methods.

tables/TableFun_sections/app/controllers/alphabet_controller.rb
```ruby
def tableView(tableView, numberOfRowsInSection: section)
  rows_for_section(section).count
end
```

tables/TableFun_sections/app/controllers/alphabet_controller.rb
```ruby
cell.textLabel.text = row_for_index_path(indexPath)

cell
```

tables/TableFun_sections/app/controllers/alphabet_controller.rb
```ruby
def tableView(tableView, didSelectRowAtIndexPath:indexPath)
  tableView.deselectRowAtIndexPath(indexPath, animated: true)

  letter = sections[indexPath.section]

  controller = UIViewController.alloc.initWithNibName(nil, bundle:nil)
  controller.view.backgroundColor = UIColor.whiteColor

  controller.title = letter

  label = UILabel.alloc.initWithFrame(CGRectZero)
  label.text = row_for_index_path(indexPath)
  label.sizeToFit
```

Finally, to get the gray headers to appear, we need to return their titles in another new dataSource method, tableView:titleForHeaderInSection::

tables/TableFun_sections/app/controllers/alphabet_controller.rb
```ruby
def tableView(tableView, titleForHeaderInSection:section)
  sections[section]
end
```

Wait a second, we've implemented the effects of changing the structure of @data, but we haven't actually changed it! If you feel compelled, you can use a list of actual names like in the earlier example on page 62, but we're going to use some Ruby tricks and just make random strings for each letter.

tables/TableFun_sections/app/controllers/alphabet_controller.rb
```ruby
@data = {}
("A".."Z").to_a.each do |letter|
  @data[letter] = []
  5.times do
    # Via http://stackoverflow.com/a/88341/910451
    random_string = (0...4).map{65.+(rand(25)).chr}.join
    @data[letter] << letter + random_string
  end
end
```

Boy, that is a lot of stuff to digest. The important part is that we set our keys of @data to be a letter and its values to be arrays of strings:

You could just as well have populated the array with letter five times, but the random string is a neat trick.

If you rake now, you can play with our gray headers.

But we're not done! In the iPod and Contacts apps, you'll notice there's a nifty scroller on the right side. This is called the *index*, and its values correspond to the section titles. Adding one to a table is really easy; we need only two new methods.

tables/TableFun_index/app/controllers/alphabet_controller.rb
```ruby
def sectionIndexTitlesForTableView(tableView)
  sections
end
def tableView(tableView, sectionForSectionIndexTitle: title, atIndex: index)
  sections.index title
end
```

sectionIndexTitlesForTableView: should return the array of strings to be shown on the index (["A", "B", "C".…"Z"]).

sectionForSectionIndexTitle: returns what section the table should jump to when a given title is tapped on the index. So, if we tap C on the index, the table view will call dataSource.tableView(self, sectionForSectionIndexTitle: "C", atIndex: 2). We could just return index, since we have a 1:1 map of sections to indices, but this is an implementation that works if that isn't true.

If you rake now, you should see our section index on the right side:

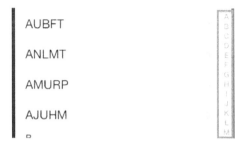

Our table is now interactive *and* user-friendly, but it's still static data. Let's make it more dynamic by allowing our alphabet-learner to delete the rows he or she doesn't find useful.

> ## Easier Sections with UILocalizedIndexedCollation
>
> Tracking section indices can be nontrivial for more complex data than just our little alphabet. Apple provides a UILocalizedIndexedCollation class that abstracts some operations we did in our example and even helps them work across multiple locales. For more information, review the Apple documentation (http://developer.apple.com/library/ios/#documentation/iPhone/Reference/UILocalizedIndexedCollation_Class/UILocalizedIndexedCollation.html) and this article on NSHipster: http://nshipster.com/uilocalizedindexedcollation/.

5.4 Swiping-to-Delete Table Rows

Swipe-to-delete is a signature iOS interaction. Notably available in Mail.app, deleting and editing data in UITableViews is a built-in API and an easy way to add some polish to your app.

To allow the user to delete table rows, you need to implement yet more dataSource methods.

- tableView:editingStyleForRowAtIndexPath:, where we determine how each row can be edited. If a row can be edited, then the table view will change the UI accordingly. After the user has acted upon that UI, like tapping the Delete button, we call commitEditingStyle:forRowAtIndexPath:.

- tableView:commitEditingStyle:forRowAtIndexPath:, which is called after the user has edited or deleted some table view element. At this point, we need to actually alter our original data structure to reflect those changes.

In our case, editingStyleForRowAtIndexPath: can be implemented pretty quickly. In practice, you may want some rows to be uneditable by using indexPath-based logic, but we're going to let every row in this app be deletable.

tables/TableFun_delete/app/controllers/alphabet_controller.rb
```
def tableView(tableView, editingStyleForRowAtIndexPath: indexPath)
  UITableViewCellEditingStyleDelete
end
```

commitEditingStyle:forRowAtIndexPath: isn't so bad either. Just use the Array delete_at() method to mutate our @data and then use UITableView's reloadData() to refresh our view.

tables/TableFun_delete/app/controllers/alphabet_controller.rb
```
def tableView(tableView,
  commitEditingStyle:editingStyle, forRowAtIndexPath:indexPath)
  if editingStyle == UITableViewCellEditingStyleDelete
    rows_for_section(indexPath.section).delete_at indexPath.row
    tableView.reloadData
  end
end
```

Let's give it a whirl, preferably with rake device because it's tricky to trigger with the simulator. It works, but it's kind of jarring how the row suddenly pops out of sight. Luckily, as with many things in iOS, we can animate it!

Animating Table Views

UITableView has some prepackaged methods for animating changes to its contents. In place of reloadData() in commitEditingStyle:, we'll use deleteRowsAtIndex-Paths:withRowAnimation:. This will automatically reload the table's data after the animation, so we don't need to worry about refreshing it ourselves. However, it means that the changes to the data source correspond precisely to animated changes, so don't try to animate removing five rows when you removed only four from your data.

tables/TableFun_animate/app/controllers/alphabet_controller.rb
```
if editingStyle == UITableViewCellEditingStyleDelete
  rows_for_section(indexPath.section).delete_at indexPath.row
  tableView.deleteRowsAtIndexPaths([indexPath],
    withRowAnimation:UITableViewRowAnimationFade)
end
```

All we need is a one-line change to really make the app look slick. There are other UITableViewRowAnimations like UITableViewRowAnimationRight or UITableViewRowAnimationBottom; always play around and choose one appropriate for your design.

5.5 Grouped-Style UITableViews

Thus far, our table cells have filled the entire width of the screen, with row after row of black-on-white Helvetica. There's actually another table *style*, one where the rows don't fill the view's frame. If you check Settings.app, you'll see what I mean.

This is UITableViewStyleGrouped, in contrast to our previous UITableViewStylePlain. Grouped table views lose the sticky headers in favor of rounded containers to denote each section. Section indexes will technically work with grouped tables, but they don't look very pleasant, so refrain from using the grouped style with very large data sets.

But wait, we haven't seen anything about table view styles, so you may be wondering where they come into play. We cheated earlier by calling UITableView#initWithFrame(). This actually isn't the designated initializer of a UITableView; instead, we should have called initWithFrame:style:.

For fun, let's see what our existing app looks like as a grouped table view. In our viewDidLoad(), just change the initializer of @table.

```
tables/TableFun_grouped/app/controllers/alphabet_controller.rb
self.title = "Alphabet"

@table = UITableView.alloc.initWithFrame(self.view.bounds,
        style: UITableViewStyleGrouped)
@table.autoresizingMask = UIViewAutoresizingFlexibleHeight
```

Let's run rake and take a look. It's a small change, but it drastically changes how our app feels, as you can see here:

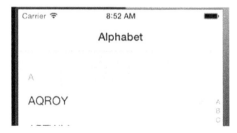

That covers the UITableView basics, which in retrospect is actually quite a lot. There is a world of customizations and behaviors to make them do really neat things, but we'll save that for another book. Even though what we created looks impressive and production-ready, it really wasn't a whole lot of code. There are several RubyMotion libraries that are trying to add more complex functionality to UITableView, such as Formotion,[2] while keeping the code just as concise. It's efforts like Formotion that really get me excited about where RubyMotion is going.

We're almost ready to make a production app, but as you can tell, our skills have progressed to a point where things can get a bit complex. It's high time we learned how to test our code to make sure it works!

---

2. https://github.com/clayallsopp/formotion

CHAPTER 6

Testing Your Apps

Automated testing is a very good idea. I won't give you the hard sell on it here (there are other books that do a better job, like *Test Driven Development* *[Bec02]*), but I will say that writing tests for your software will save hours upon hours of time as your product grows more complex. If your current code doesn't work as intended or something breaks in the future, you'll know it immediately. Sounds like something we should really use in our iOS apps, right?

The APIs we've covered so far are essentially common to iOS development in both Objective-C and RubyMotion. The syntax may be a little different here and there, but all the classes and methods are part of the Apple's Cocoa frameworks. Testing, however, is a different world in RubyMotion.

The Ruby community has taken automated testing to heart. RubyMotion ships with a variant of the popular RSpec framework, a concise library that makes authoring new tests painless and requires little extra effort on the programmer's part. In contrast to the brevity of Ruby, the Apple method of automated testing involves writing JavaScript or very verbose Objective-C. Let's look at just how easy testing iOS apps can be with idiomatic Ruby.

6.1 Constructing Basic Tests

The simplest tests are designed to test just one component of our code at a time, independent of how it interacts with other classes or the global state of our app. Tests consist of simple *assertions* such as object equality or existence. If any of our assertions fail, then the entire test fails and stops executing. We're going to write some simple tests for an app and go through some common use cases you might run into.

Let's create a new RubyMotion project with `motion create TestFun`. Remember way back in *Creating a New App* when we discussed the default files and folders motion create generates? Well, we're going to take a look at one folder we previously neglected: spec.

RubyMotion reads your tests from the spec folder, recursively loading every single .rb file it contains. One way to organize your tests is to make each file correspond to a class in your code.

When you create a RubyMotion app, it generates a default test in spec/main_spec.rb that looks like this:

```
testing/TestFun/spec/main_spec.rb
describe "Application 'TestFun'" do
  before do
    @app = UIApplication.sharedApplication
  end
  it "has one window" do
    @app.windows.size.should == 1
  end
end
```

Sit there for a moment and imagine every word of that being read out loud. See how expressive and simple that is? @app.windows.size.should == 1 does exactly what it sounds like: if size is not equal to 1, then the test fails.

should() is a helper method added to every object in the testing environment and forms the basis of most tests. It uses the default meaning of the == operator for each object type, so it can be used for more than just integers. Here are some more examples:

```
@user.nil?.should == false
```

```
@model.id.should == 5
```

```
[1,2,3].should.not == [1,2,3,4]
```

The last example uses the not() helper method, which negates the intention of should(). We could have just used should() with an additional ! on the right side, but chaining a not() method improves readability.

The generated test in main_spec also has the describe() and it() blocks, which add structure to our tests. You can read the combination of the two blocks like a sentence: "Test that the application 'TestFun' has one window." A describe() block can have many it() blocks, and an it() block can have many assertions. You can even nest describe()s to create very granular descriptions of each test. We'll see the results of all this structuring in just a second when we run the tests.

The last element of the generated test is the before() block. The code in a before() block runs prior to every sibling test block. It's a good place to reset your objects and restore the state of the code you're testing. Similarly, you can define an after() block to execute after every test.

So, how do we actually use all these fun toys? In your terminal window, run rake spec. This will build your app and run the specs against it. In addition to the app briefly appearing in the simulator, you should see some output in the terminal like this:

```
$ rake spec
   Build ./build/iPhoneSimulator-7.1-Development
 ...
 Simulate ./build/iPhoneSimulator-7.1-Development/TestFun_spec.app

Application 'TestFun'
  - has one window [FAILED]

Bacon::Error: 0.==(1) failed
  spec.rb:553:in `satisfy:': Application 'TestFun' - has one window
  ...

1 specifications (1 requirements), 1 failures, 0 errors
```

RubyMotion will run all of our tests and yell if something goes wrong. You can see here how the test output reflects the describe() blocks ("Application 'TestFun'") and the it() blocks within them ("- has one window").

The generated test actually fails! The lone assertion that we have a UIWindow fails, as indicated by the [FAILED] beside the test description and the summary at the bottom. One current headache in RubyMotion is that the stack trace does not reveal the line number in our original main_spec; thankfully, we can deduce it from the spec description and assertion type.

Let's fix this test. In our AppDelegate, we add a UIWindow just like in all of our previous apps.

testing/TestFun_passing/app/app_delegate.rb
```ruby
class AppDelegate
  def application(application, didFinishLaunchingWithOptions:launchOptions)
    @window = UIWindow.alloc.initWithFrame(UIScreen.mainScreen.bounds)
    @window.makeKeyAndVisible
    true
  end
end
```

Run rake spec again and give yourself a pat on the back.

```
$ rake spec
...
1 specifications (1 requirements), 0 failures, 0 errors
```

You can actually get pretty far just by adding new test files to spec the few RSpec methods we've covered (such as describe() and should()). Tests involving models or algorithmic code should be completely expressible with even these basic methods. But unfortunately, iOS apps aren't just static bits of code: we often need to test our callback-centric user interface.

6.2 Testing App UI and Controllers

In addition to a basic RSpec testing framework, RubyMotion includes extensions that help us test complex interactions with views and a device. You can specifically test events such as taps, flicks, and pinches to ensure that your callbacks are functioning accordingly.

Let's walk through an example of RubyMotion's UI test helpers. We're going to programmatically tap a UIButton and assert that some state has changed after the tap. Even better, we'll make sure to visibly alter the UI after the tap to make it obvious while the test is running in the simulator.

Since UI testing requires a UIViewController instance, we need one of our own. Create the app/controllers directory and a button_controller.rb file within. Our implementation doesn't have anything surprising; we're just creating a button in viewDidLoad().

testing/TestFun_ui/app/controllers/button_controller.rb
```ruby
class ButtonController < UIViewController
  def viewDidLoad
    super
    button = UIButton.buttonWithType(UIButtonTypeSystem)
    button.setTitle("Tap Me", forState:UIControlStateNormal)
    button.sizeToFit
    button_origin = [0, 100]
    button.frame = [button_origin, button.frame.size]
    button.addTarget(self,
      action:'tapped',
      forControlEvents:UIControlEventTouchUpInside)
    self.view.addSubview(button)
```

After tapping, we need to change some variable of ButtonController. Let's add a quick and dirty instance variable before finishing viewDidLoad().

testing/TestFun_ui/app/controllers/button_controller.rb
```ruby
    self.view.addSubview(button)

    @tapped = false
```

The button callback will not only change @tapped but change the view's background color as well. When the specs run, we'll see the app flash colors and know for sure that the RubyMotion UI helpers are working as intended.

testing/TestFun_ui/app/controllers/button_controller.rb
```
def tapped
  @tapped = true
  self.view.backgroundColor = UIColor.redColor
end
```

Finally, we need to hook up the controller in our AppDelegate. Simply set a new instance as the rootViewController, give it a rake, and make sure it looks like Figure 8, *Preparing our UI test*, on page 73 after tapping the button.

testing/TestFun_ui/app/app_delegate.rb
```
@window = UIWindow.alloc.initWithFrame(UIScreen.mainScreen.bounds)
@window.backgroundColor = UIColor.whiteColor
@window.rootViewController = ButtonController.alloc.initWithNibName(nil, bundle:nil)
@window.makeKeyAndVisible
```

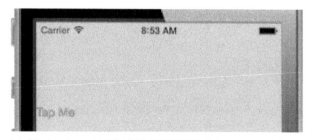

Figure 8—Preparing our UI test

Now that our UI is ready, it's time to write the actual test. Since we're keeping our project organized, create spec/button_controller_spec.rb to store our test. Let's look at how we implement this, since it's quite short.

testing/TestFun_ui/spec/button_controller_spec.rb
```
describe "ButtonController" do
  tests ButtonController
  it "changes color after tapping" do
    tap("Tap Me")
    controller.instance_variable_get("@tapped").should == true
  end
end
```

All that earlier code for these few lines? Indeed, UI testing is just as expressive as more basic tests. Each UI test focuses on a single UIViewController and will actually instantiate a new UIWindow containing one isolated controller. The tests() method after describe() is what triggers this behavior; simply pass it the appropriate class, and RubyMotion handles the rest.

By declaring our test to be a UI test, we get some new helper methods. Here we use tap(), which takes either an instance of UIView or an *accessibility label*. Accessibility labels are used by the iOS VoiceOver utility to read-aloud parts of the screen. You can set these labels yourself by changing the accessibilityLabel property of view, but some classes can derive their label value from other properties such as title. UIButton is one such class, so we can safely pass "Tap Me" to tap().

If we run rake spec, we'll see the test run in the simulator and the background momentarily flash red. We can confirm in the console that our ButtonController test does pass, but something has gone wrong in our original test!

```
$ rake spec
...

ButtonController
  - changes color after tapping

Application 'TestFun_ui'
  - has one window [FAILED]


  ...
2 specifications (2 requirements), 1 failures, 0 errors
```

We briefly went over how RubyMotion runs each controller test in isolation by creating a new UIWindow; thus, our test that there is only one window should indeed fail. If you want, you can rewrite the test to confirm that we have a rootViewController instead.

testing/TestFun_ui/spec/main_spec.rb
```
it "has one controller" do
  controller = @app.keyWindow.rootViewController
  controller.is_a?(ButtonController).should == true
end
```

That wraps it up for our little testing introduction. We've gone over the skills to write solid tests for your apps, but it is up to you to put them in action. For detailed information on the testing framework, check out the RubyMotion Developer Center's article on testing.[1]

Testing is one topic where I think RubyMotion really shines. Objective-C testing frameworks are nontrivial to set up in Xcode, and Apple's prescribed

---

1. http://www.rubymotion.com/developer-center/articles/testing/

testing solution isn't nearly as clean as RubyMotion's RSpec-like format. Plus, since tests are just a call to rake spec away, it's very easy to integrate Ruby-Motion into a continuous integration environment.

Now that we can build and test the typical iOS UI, it's time to put it all together and hook it up to the Internet. In *Example: Writing an API-Driven App*, we're going to create a small example app that gets and sends data via an HTTP API.

CHAPTER 7

Example: Writing an API-Driven App

It's time for the main event. We're going to put together *everything* we know about iOS development and write a complete app. This chapter will introduce only one topic we haven't covered yet: sending and receiving information from the Internet via HTTP. Nowadays most software needs to communicate to an external service in some way, so this example is highly relevant to whatever you try to build. Since this isn't a book on iOS design, we might not wind up with the prettiest software on a device, but it will do everything we need.

We're going to build a mobile app for the Colr JSON API (http://www.colr.org/ api.html). Our users can type in a color hex code (i.e., #3B5998) and then see what tags Colr users have assigned to that color. If our user is feeling particularly adventurous, they can even add a new tag!

Let's talk about high-level architecture. When building a production app, it's a good idea to break down all of the objects and classes we need to create. The app will have two controllers: a SearchController for searching and a ColorController to display detailed information about a color. These should be wrapped inside a UINavigationController, our top-level controller. We're also going to need some models for our data: Color and Tag.

Now that we have a good idea of what code to write, let's set up our project.

7.1 Setting Up the Project

Create our RubyMotion project with motion create Colr. To keep our project organized, add the following files and their directories:

- ./app/controllers/search_controller.rb
- ./app/controllers/color_controller.rb
- ./app/models/color.rb
- ./app/models/tag.rb

It already feels like a big fancy app, doesn't it? Now we need to add Bub-bleWrap to our project. Earlier we used BubbleWrap to clean up key-value observing, but it also includes pleasant wrappers for HTTP requests. Refer to *Changing Models with Key-Value Observing* for detailed instructions, but essentially what we do is install the bubble-wrap gem and require it in our Rakefile.

Now that we have the pieces in place, it's time to start writing some code.

7.2 Adding Hash-Friendly Models

Let's start with the models. The model code should be independent of any other classes, so filling them in is usually a good place to start. Resist the temptation to make explicit references to your views or controllers!

First up are the Color objects. The Colr API returns its colors as JSON objects in the following form:

```
{
  "timestamp": 1285886579,
  "hex": "ff00ff",
  "id": 3976,
  "tags": [{
    "timestamp": 1108110851,
    "id": 2583,
    "name": "fuchsia"
  }]
}
```

Right away, it looks like we need timestamp, hex, id, and tags properties. In particular, tags represents a has-many relationship with Tag objects.

Let's get down to business. Open color.rb and start filling it in with our hash-based model template from Section 4.2, *Preparing Scalable Models*, on page 46.

api_example/Colr/app/models/color.rb
```
class Color
  PROPERTIES = [:timestamp, :hex, :id, :tags]
  attr_accessor *PROPERTIES
  def initialize(hash = {})
    hash.each { |key, value|
      if PROPERTIES.member? key.to_sym
        self.send((key.to_s + "=").to_s, value)
      end
    }
  end
```

Now we need to fix tags to be an array and coerce its contents into Tag objects. You should finish the definition of Color like this:

api_example/Colr/app/models/color.rb
```ruby
  def tags
    @tags ||= []
  end
  def tags=(tags)
    if tags.first.is_a? Hash
      tags = tags.collect { |tag| Tag.new(tag) }
    end
    tags.each { |tag|
      if not tag.is_a? Tag
        raise "Wrong class for attempted tag #{tag.inspect}"
      end
    }
    @tags = tags
  end
end
```

We override #tags() to guarantee that it returns an Array even if no values have been given. This makes logical sense given the has-many relationship and allows us to not worry about checking for nil elsewhere.

The tags=() setter ensures that every object in tags will be a Tag object if possible. We could just accept an array of hashes straight from the JSON, but then we lose the ability to perform straightforward KVO on those objects. Since we're on the subject, it's a good idea to start working on the Tag implementation.

As you saw in the JSON for colors, the API returns tags in the following form:

```json
{
  "timestamp": 1108110851,
  "id": 2583,
  "name": "fuchsia"
}
```

We'll need timestamp, id, and name properties but no special relationships this time. In tag.rb, just use the scalable model template with those attributes to create our Tag class.

api_example/Colr/app/models/tag.rb
```ruby
class Tag
  PROPERTIES = [:timestamp, :id, :name]
  attr_accessor *PROPERTIES
  def initialize(hash = {})
    hash.each { |key, value|
      if PROPERTIES.member? key.to_sym
        self.send((key.to_s + "=").to_s, value)
      end
    }
  end
end
```

That's all for our models. Pretty painless, right? Using these hash-friendly implementations lets us easily instantiate objects using the server's JSON. Now on to the meat of our app, the controllers.

7.3 Making GET Requests and SearchController

The first controller the user interacts with is the SearchController. Here, the user can enter a hexadecimal color code like #3B5998 and tap a button to see detailed information about that color. The only edge case we need to worry about is if the Colr database doesn't contain that color; in that case, we'll alert the user accordingly.

Setting Up the UI

How does the user enter text? We're going reach back all the way to Section 2.6, *Making Text Dynamic with UITextField*, on page 21 and use the venerable UITextField. We'll use it along with a UIButton to actually trigger the search.

Let's start setting up these elements. Open search_controller.rb and fill in viewDid-Load() with our new subviews.

api_example/Colr/app/controllers/search_controller.rb

```
class SearchController < UIViewController
  def viewDidLoad
    super
    self.title = "Search"
    self.view.backgroundColor = UIColor.whiteColor

    @text_field = UITextField.alloc.initWithFrame [[0,0], [160, 26]]
    @text_field.placeholder = "#abcabc"
    @text_field.textAlignment = UITextAlignmentCenter
    @text_field.autocapitalizationType = UITextAutocapitalizationTypeNone
    @text_field.borderStyle = UITextBorderStyleRoundedRect
    @text_field.center = [
        self.view.frame.size.width / 2,
        self.view.frame.size.height / 2 - 100]
    self.view.addSubview(@text_field)

    @search = UIButton.buttonWithType(UIButtonTypeSystem)
    @search.setTitle("Search", forState:UIControlStateNormal)
    @search.setTitle("Loading", forState:UIControlStateDisabled)
    @search.sizeToFit
    @search.center = [
        self.view.frame.size.width / 2,
        @text_field.center.y + 40]
    self.view.addSubview(@search)
  end
end
```

Most of this is just specific positioning and sizing stuff that's old-hat by now. One new bit is UIControlStateDisabled, which corresponds to what a control looks like if we set enabled to false. UITextBorderStyleRoundedRect is one style to make UITextFields look nice without any additional configuration, but StyleNone and StyleLine also exist.

Before we run our app, we need to set up our AppDelegate to use this controller. As we've done many times before, create a UINavigationController with our Search-Controller at its root.

api_example/Colr/app/app_delegate.rb
```
class AppDelegate
  def application(application, didFinishLaunchingWithOptions:launchOptions)
    @window = UIWindow.alloc.initWithFrame(UIScreen.mainScreen.bounds)
    @window.backgroundColor = UIColor.whiteColor

    @search_controller = SearchController.alloc.initWithNibName(nil, bundle:nil)
    @navigation_controller =
      UINavigationController.alloc.initWithRootViewController(@search_controller)

    @window.rootViewController = @navigation_controller
    @window.makeKeyAndVisible
    true
  end
end
```

At this point, our Colr app should look like this:

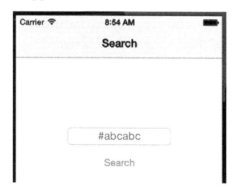

Sadly, it doesn't do a whole lot yet. Let's start filling in that implementation with some callbacks.

Running GET Requests

Up to this point, we've detected UIButton taps with addTarget:action:forControlEvents:. It works, but having an external callback method passed by name is an awfully un-Ruby thing to do. BubbleWrap has a nifty little when() wrapper to pass detect control events in a block, which makes our code much more readable. In viewDidLoad(), add one of these blocks for our @search button:

api_example/Colr_search/app/controllers/search_controller.rb
```
self.view.addSubview(@search)

@search.when(UIControlEventTouchUpInside) do
  @search.enabled = false
  @text_field.enabled = false

  hex = @text_field.text
  # chop off any leading #s
  hex = hex[1..-1] if hex[0] == "#"

  Color.find(hex) do |color|
    @search.enabled = true
    @text_field.enabled = true
  end
end
```

The when() function is available to every UIControl (of which UIButton is a subclass) and takes the usual bitmask of UIControlEvents as its arguments. While the request runs, we temporarily disable our UI elements using the enabled property of each element.

But wait, where did that Color.find() method come from? You should keep all of your URL requests inside of models instead of controllers. It keeps your controllers leaner, and if we want to grab a Color from the server somewhere else in the app, then there's no copy-paste code duplication. And who knows, maybe we'll need to do that too...*foreshadowing*.

It's time to add that find() static method to our Color. We're *finally* going to play with HTTP requests here, starting with HTTP.get.

api_example/Colr_search/app/models/color.rb
```
def self.find(hex, &block)
  BubbleWrap::HTTP.get("http://www.colr.org/json/color/#{hex}") do |response|
    p response.body.to_str
    block.call(nil)
  end
end
```

Not too bad, right? We use HTTP.get to send a GET request to the server via the correct API URL. Note that we use the block argument to make it plain that this method is intended to be used with a block. If you're fuzzy on this part of Ruby, it means block isn't explicitly passed as another argument but rather implicitly when we add a do/end after calling the method. The number and order of variables in block.call(some, variables) correspond to their representations in do |some, variables|.

Go ahead and rake and test it with a color like ff00ff. You should see something like this output in the console:

```
(main)> "{\"colors\": [
  {\"timestamp\": 1285886579, \"hex\": \"ff00ff\", \"id\": 3976,
    \"tags\": [
      {\"timestamp\": 1108110851, \"id\": 2583, \"name\": \"fuchsia\"},
      {\"timestamp\": 1108110864, \"id\": 3810, \"name\": \"magenta\"},
      {\"timestamp\": 1108110870, \"id\": 4166, \"name\": \"magic\"},
      {\"timestamp\": 1108110851, \"id\": 2626, \"name\": \"pink\"},
      {\"timestamp\": 1240447803, \"id\": 24479, \"name\": \"rgba8b24ff00ff\"},
      {\"timestamp\": 1108110864, \"id\": 3810, \"name\": \"magenta\"}
  ]}]
  ...
}"
```

That looks an awful lot like JSON, doesn't it? That makes sense, given the /json/ in the URL. Wouldn't it be convenient if we could parse that into a normal Ruby hash and use it to instantiate our hash-friendly models?

BubbleWrap to the rescue again! Our nifty friend has a BubbleWrap::JSON.parse() method, which takes a JSON string and spits out a Ruby hash. Let's update Color.find() to use it.

api_example/Colr_search2/app/models/color.rb
```ruby
def self.find(hex, &block)
  BubbleWrap::HTTP.get("http://www.colr.org/json/color/#{hex}") do |response|
    result_data = BubbleWrap::JSON.parse(response.body.to_str)
    color_data = result_data["colors"][0]

    # Colr will return a color with id == -1 if no color was found

    color = Color.new(color_data)
    if color.id.to_i == -1
      block.call(nil)
    else
      block.call(color)
    end

  end
end
```

Nice and easy. Our edge-case comes into play here when the only color returned has an id of -1, meaning no results were found. We need to update our SearchController to deal with that and the normal, successful case.

api_example/Colr_search2/app/controllers/search_controller.rb

```
Color.find(hex) do |color|
➤    if color.nil?
➤      @search.setTitle("None :(", forState: UIControlStateNormal)
➤    else
➤      @search.setTitle("Search", forState: UIControlStateNormal)
➤      self.open_color(color)
➤    end
     @search.enabled = true
     @text_field.enabled = true
   end
```

api_example/Colr_search2/app/controllers/search_controller.rb

```
➤ def open_color(color)
➤   p "Opening #{color.inspect}"
➤ end
```

For now, we just print out the returned Color object upon a successful search. If you run the app now and search a color, the console should display all the properties of the Color in addition to the related Tag objects. Pretty neat, right? But now we need to change open_color(color) to its final implementation with a ColorController class. I guess now is a good time to create that controller.

7.4   POST Requests and ColorController

ColorController will be a bit more involved than a text field and one button. The controller's view will have two sections: a lower-half UITableView to list the color's Tags and an upper section with an entry field to add new tags.

When we want to add a new tag, we send a POST request to Colr with the tag's text. When that request finishes, we completely refresh our data and update the UI so the user knows the new tag is saved. POST requests are done just like GETs, except using HTTP.post. That sounds like a lot, so let's take it one step at a time.

Setting Up the UI

First we're going to use a custom initializer for ColorController. This will take a Color as its sole argument, which makes sense given that the controller should always be associated with a color. In color_controller.rb, start our class like this:

api_example/Colr_detail/app/controllers/color_controller.rb

```
class ColorController < UIViewController
  attr_accessor :color
  def initWithColor(color)
    initWithNibName(nil, bundle:nil)
    self.color = color
    self
  end
```

As we covered in Section 3.1, *Adding a New UIViewController*, on page 27, when writing an iOS initializer, you need to do two things: call the designated initializer and return self at the end. We'll store the color using an instance variable @color, which attr_accessor() uses to create the corresponding getter and setter.

Next we need to lay out the interface. We're going to split the viewDidLoad() method into two parts: one for the upper add-tag section and one for the tags table view. Brace yourself—lots of frame code is coming.

api_example/Colr_detail/app/controllers/color_controller.rb

```ruby
def viewDidLoad
  super
  self.title = self.color.hex
  self.edgesForExtendedLayout = UIRectEdgeNone
  self.view.backgroundColor = UIColor.whiteColor

  padding = 10

  @info_container = UIView.alloc.initWithFrame(
    [[0, 0], [self.view.frame.size.width, 60]])
  @info_container.backgroundColor = UIColor.lightGrayColor
  self.view.addSubview(@info_container)

  box_size = @info_container.frame.size.height - 2*padding
  @color_view =
    UIView.alloc.initWithFrame([[padding, padding], [box_size, box_size]])
  @color_view.backgroundColor = String.new(self.color.hex).to_color
  self.view.addSubview(@color_view)

  text_field_origin = [
    @color_view.frame.origin.x + @color_view.frame.size.width + padding,
    @color_view.frame.origin.y]
  @text_field = UITextField.alloc.initWithFrame(CGRectZero)
  @text_field.placeholder = "tag"
  @text_field.autocapitalizationType = UITextAutocapitalizationTypeNone
  @text_field.borderStyle = UITextBorderStyleRoundedRect
  @text_field.contentVerticalAlignment = UIControlContentVerticalAlignmentCenter
  self.view.addSubview(@text_field)

  @add = UIButton.buttonWithType(UIButtonTypeSystem)
  @add.setTitle("Add", forState:UIControlStateNormal)
  @add.setTitle("Adding...", forState:UIControlStateDisabled)
  @add.setTitleColor(UIColor.lightGrayColor, forState:UIControlStateDisabled)
  @add.sizeToFit
  @add.frame = [
    [self.view.frame.size.width - @add.frame.size.width - padding,
      @color_view.frame.origin.y],
    [@add.frame.size.width, @color_view.frame.size.height]]
```

```
self.view.addSubview(@add)
add_button_offset = @add.frame.size.width + 2*padding
@text_field.frame = [
  text_field_origin,
  [self.view.frame.size.width - text_field_origin[0] - add_button_offset,
  @color_view.frame.size.height]]
```

Whew, that is a lot of code, isn't it? This is where Interface Builder comes in handy and reduces a lot of the complex calculations, like setting the frame of @text_field only after @add is completed. At this point, I think we know enough about how UIViews work for you to make your own choice about when to use IB and when to do it all in code.

Now that we have some our UI set up, let's take it for a spin. We need to go back to SearchController and fix that open_color() to use a ColorController.

api_example/Colr_detail/app/controllers/search_controller.rb
```
def open_color(color)
➤   controller = ColorController.alloc.initWithColor(color)
➤   self.navigationController.pushViewController(controller, animated:true)
end
```

If we rake, you should see a nice upper bar, like so:

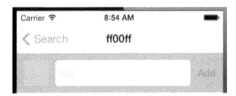

Just imagine that we have some slick gradients and 1-pixel shadows.

Next we need to add our UITableView of tags. After all of our other view creation, end viewDidLoad() by creating the table view and making the controller its dataSource.

api_example/Colr_detail2/app/controllers/color_controller.rb
```
  table_height = self.view.bounds.size.height - @info_container.frame.size.height
  table_frame = [[0, @info_container.frame.size.height],
    [self.view.bounds.size.width, table_height]]
  @table_view =UITableView.alloc.initWithFrame(table_frame,
    style: UITableViewStylePlain)
  @table_view.autoresizingMask = UIViewAutoresizingFlexibleHeight
  self.view.addSubview(@table_view)
  @table_view.dataSource = self
end
```

Since we're now done with the dataSource, we need to add the requisite methods to the controller. And since we already have the array we need with self.color.tags, our methods can be pretty lean.

```
api_example/Colr_detail2/app/controllers/color_controller.rb
def tableView(tableView, numberOfRowsInSection:section)
  self.color.tags.count
end
def tableView(tableView, cellForRowAtIndexPath:indexPath)
  @reuseIdentifier ||= "CELL_IDENTIFIER"
  cell = tableView.dequeueReusableCellWithIdentifier(@reuseIdentifier)
  cell ||=
    UITableViewCell.alloc.initWithStyle(
      UITableViewCellStyleDefault, reuseIdentifier:@reuseIdentifier)
  cell.textLabel.text = self.color.tags[indexPath.row].name
  cell
end
```

I told you it wasn't very hard, didn't I? Let's try rake again and see how our tags look.

Now we get to the fun part: running POST requests to add *new* tags to a color.

Running POST Requests

Now that our UI is complete, it's time to implement adding new tags. Just like we created a Color.find() method, we're going to add a Color#add_tag() method to keep our API requests out of the controllers.

Let's add our new method in color.rb. All of BubbleWrap's HTTP methods allow a hash payload argument. For GET requests, this correctly appends the hash's keys and values as a URL query; for POST/PUT/DELETE, they are added into the request body. Our implementation should look something like this:

```
api_example/Colr_detail3/app/models/color.rb
def add_tag(tag, &block)
  BubbleWrap::HTTP.post("http://www.colr.org/js/color/#{self.hex}/addtag/",
    payload:{tags: tag}) do |response|
      if response.ok?
        block.call(tag)
      else
        block.call(nil)
      end
  end
end
```

Just like find(), we use the block argument for callbacks. However, this time we use the ok?() method of the response to detect whether everything worked, which checks for non-200 status codes.

Our block takes one argument, returning nil if the request fails. Let's hook this up into the callback for the @add button in ColorController.

api_example/Colr_detail3/app/controllers/color_controller.rb
```ruby
self.view.addSubview(@add)

@add.when(UIControlEventTouchUpInside) do
  @add.enabled = false
  @text_field.enabled = false
  self.color.add_tag(@text_field.text) do |tag|
    if tag
      refresh
    else
      @add.enabled = true
      @text_field.enabled = true
      @text_field.text = "Failed :("
    end
  end
end
```

Not too bad, right? We remember to disable our UI while the request runs, which prevents the user from entering multiple new tags and allowing for weird race conditions. If add_tag() works, we call this refresh() method that we haven't implemented yet. refresh() will sync our Color to the current server data and then refresh the tags table so we see that everything worked. We can reuse our Color.find() method, which looks pretty handy right now, doesn't it?

api_example/Colr_detail3/app/controllers/color_controller.rb
```ruby
def refresh
  Color.find(self.color.hex) do |color|
    self.color = color
    @table_view.reloadData

    @add.enabled = true
    @text_field.enabled = true
  end
end
```

Fantastic, we're done! One topic we didn't discuss was how to test this app. RubyMotion's framework gives you the building blocks for testing basic code, but testing dynamic features like HTTP requests makes things tricky. Thankfully, there are third-party libraries that can help you quickly test these more complex components. For example, WebStub (https://github.com/mattgreen/webstub) is a RubyMotion gem that allows you to transparently configure the data returned from any URLs your app requests. This is especially useful if you want to confirm the behavior for when things go wrong and erroneous data is returned from the server.

Our little Colr app may not have used the most popular API, but it's a good example of how to structure any API-based app. And even if you aren't interacting with a complex JSON API, BubbleWrap's HTTP wrapper might be useful for more generic requests.

Now that you've built an app from start to more or less finish, it's time to ship the final product to your users. Let's dig into how we prepare and submit our creation to the App Store.

CHAPTER 8

Uploading to the App Store

We've come a long way together. We started with nothing, and now we're well on our way to making beautiful production-ready iOS apps. That wasn't so hard, right? At this point, you could go off and continue your iOS adventure, but at the end of the day, there's still one more skill we need to cover: uploading our apps to Apple's App Store.

To get your product on the App Store, you'll need to be a paying member of the iOS Developer Program (https://developer.apple.com/programs/ios/). So far, we've just built our apps for the iOS simulator and everything worked out, but now we need real developer certificates to sign our apps. You should follow the instructions on Apple's website; when you're done setting up your account, we're ready to move on.

8.1 Including App Resources

So far, we've just been shipping code with our apps, but most projects require a variety of resources: images, sounds, or just raw data. RubyMotion has pretty seamless methods for including each of these, so let's take look.

Bundling Icons

Dribbble-worthy or not, your app's icon will appear in many different locations and on varying resolutions: iPhones, iPads, iTunes, and maybe more in the future. To preserve the icon's visual integrity, Apple encourages developers to ship multiple versions of the icon with every app.

To include your icon files, simply add them to the resources folder. They can have any name, but being descriptive is never a bad idea (such as Icon-72 and Icon-144). Finally, set the icons() array of our app in the Rakefile to contain these filenames:

```
Motion::Project::App.setup do |app|
  # ...
  app.icons =  ["Icon-72.png", "Icon-144.png", ...]
end
```

Upon your next rake, the simulator's home screen should be updated with the proper icon.

By default, iOS will apply a curved shine effect on top of your icon, just like the icons for iTunes and the App Store. This might be desired behavior, but if it's not, then it's easy to disable. Simply set the prerendered_icon flag to true in the setup() block.

```
Motion::Project::App.setup do |app|
  # ...
  app.prerendered_icon = true
end
```

That's about it for RubyMotion and app icons. For a complete list of resolutions and more design advice, refer to Apple's Human Interface Guidelines at the following URL:

http://developer.apple.com/library/ios/#documentation/userexperience/conceptual/mobilehig/IconsImages/IconsImages.html

File Resources

There are more resources than just icons. Most apps have customized interfaces or graphics, and we need to load those from images; you may even need to use specific fonts or some sort of raw data. RubyMotion provides an elegant solution for bundling these files with your app: simply put them in the resources folder.

Any files included in resources will be bundled with your app and accessible in your code. For example, adding a face.png file allows us to create a UIImage instance with UIImage.imageNamed("face"). All supported .ttf and .otf fonts in resources will be included properly in your app and allow you to use the UIFont.fontWithName("MyFont", size:20) method.

If you want to grab the contents of an arbitrary file like data.txt, we need to grab its path.

```
path = NSBundle.mainBundle.pathForResource("data", ofType:"txt")
data = NSData.dataWithContentsOfFile(path)
```

You can then transform the NSData instance as necessary.

Configuration Properties with Info.plist

Want to know a secret? Most of the app properties in the Rakefile and displayed with rake config are actually stored in one central data structure: Info.plist. RubyMotion generates this file behind the scenes using those nifty helper methods; however, the app settings are not exhaustive of the possible properties. Occasionally, we may need to edit Info.plist manually.

Info.plist stores information in a Hash-like structure, which is how RubyMotion gives us more direct access to the file. For example, if we wanted to assign our app a custom URL scheme, we would edit the CFBundleURLTypes property like so:

```
Motion::Project::App.setup do |app|
  # ...
  app.info_plist['CFBundleURLTypes'] = [
    {
      'CFBundleURLName' => 'com.mycomapny.myapp',
      'CFBundleURLSchemes' => ['myapp']
    }
  ]
end
```

With that, our app will now open if the device is opened to a *myapp://* URL. You can also output app.plist() to view or debug the complete list of properties.

For a complete list of valid properties in Info.plist, consult Apple's documentation at the following URL:

https://developer.apple.com/library/mac/#documentation/General/Reference/InfoPlistKeyReference/Introduction/Introduction.html#//apple_ref/doc/uid/TP40009247

8.2 Archiving for Release

Now that your app is coded, tested, and configured to your liking, it's time to ship. Again, you'll need to be registered with Apple's Developer Program before attempting to upload an app. You can find instructions for installing your code-signing certificate and provisioning profile in the iOS Team Administration guide at the following URL:

http://developer.apple.com/library/ios/#documentation/ToolsLanguages/Conceptual/DevPortalGuide/CreatingandDownloadingDevelopmentProvisioningProfiles/CreatingandDownloadingDevelopment-ProvisioningProfiles.html

Once that's done, we're good to go.

Before we finish the remaining settings in RubyMotion, you should add your new app or update in iTunes Connect (http://itunesconnect.apple.com). iTunes

Connect is Apple's developer front end to the App Store; before uploading your files, you need to complete all of the required information. Apple has a very thorough guide on using iTunes Connect that you can consult if you run into issues (at the following URL).

http://developer.apple.com/library/ios/#documentation/LanguagesUtilities/Conceptual/iTunesConnect_Guide/1_Introduction/Introduction.html

SDK Version and Deployment Target

In your Rakefile, you should set what iOS SDK version you're building against and what iOS versions you'll allow the app to run on. By default, RubyMotion will use the latest and greatest SDK version for everything, but sometimes you may want to make your app available to older devices.

You can set the app's sdk_version property to the stringified iOS version, such as "5.0". This will be the SDK your app is compiled against; in other words, your app will fail to build if you use APIs from newer SDKs. Apple usually only takes apps built against the most two most recent versions, so building against something like iOS 4.3 isn't recommended.

On the other hand, deployment_target sets the lowest supported iOS version that our app can run on. But wait, how do we let our app run on operating systems older than what we compile against? In those cases, you need to check for the availability of classes at runtime using respond_to?() and const_get().

For example, if we wanted to use some features of iOS6 such as Facebook integration but gracefully fall back to Twitter on older devices, our Rakefile would look like this:

```
Motion::Project::App.setup do |app|
  # ...
  app.sdk_version = "6.0"
  app.deployment_target = "5.0"
end
```

Required Configuration

Before we build our app, there are some options that need to be set. They aren't necessary while we debug our app, but they are required before submitting to the App Store.

Your app's version is the string displayed on your app's App Store page and when listing available updates. It must consist of numbers and decimals, and each successive update must contain a higher version number. For example, "1.1" and "1.2.3" are valid versions; "1.3a" and "cat" are not.

The identifier is also required. When you create your app in iTunes Connect, you are required to create an App ID. These are created in the iOS Provisioning Portal and follow a reverse domain name format like "com.mycompany.myappname". RubyMotion provides a temporary default during development, but Apple will require a permanent unique ID when you submit your app.

Lastly, you need to set the provisioning profile used for releasing your app. provisioning_profile should be the complete path to the distribution profile you created in the iOS Provisioning Portal (if you haven't, now is the time). The default location to install these is ~/Library/MobileDevice/Provisioning Profiles, but it's fine if they're placed anywhere. When testing on the simulator, RubyMotion uses the first development certificate it finds in that default location; however, building an app for the App Store requires you to use a specific distribution profile.

A robust App Store–ready Rakefile should look something like this:

```
Motion::Project::App.setup do |app|
  # ...
  app.sdk_version = "6.0"
  app.deployment_target = "5.1"
  app.version = "1.0"
  app.identifier = "com.myappcompany.myawesomeapp"
  app.provisiong_profile = "/Users/me/profiles/1234.mobileprovision"
end
```

Development vs. Release

Most software projects, including RubyMotion, have at least two build configurations: development and release. Development builds can include all kinds of debugging statements or even features that we don't want to ship in the final product, while release builds should be safely isolated from any harmful or verbose code.

By default, running rake simulator or device will build the app in development mode; archiving the app for the App Store will build it in release mode. If you want to change your app build settings on a per-mode basis, you can use the development() and release() methods like so:

```
Motion::Project::App.setup do |app|
  # ...
  app.development do
    # code runs in dev mode
    app.version = "1.0.1"
    app.identifier = "com.myappcompany.myawesomeapp.dev"
  end
  app.release do
```

```
    # code runs in release mode
    app.version = "1.0"
  end
end
```

For a complete list of available options and more information, check out the RubyMotion project management documentation (http://www.rubymotion.com/developer-center/guides/project-management/).

Once everything is in place, the magic command is rake archive:distribution. This will build your app and place the final product in ./build/iPhoneOS-SDK_VERSION-Release/APP_NAME.ipa. This *archive* contains your application binary and resources and is signed using the distribution certificate you created when signing up for a developer account.

Now that we have our finished .ipa, we can use the Application Loader, which ships with Xcode to upload our app.

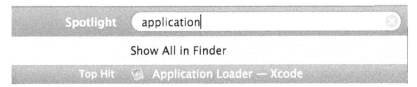

Before you upload the archive, you'll need to have your app data added on iTunes Connect. Once the status of your app on iTunes Connect is "Waiting for Upload," follow the instructions given in Application Loader to complete the process. After that, all you have to do is wait for Apple to approve your app!

8.3 What's Next?

And that's all she wrote, and the fat lady has sung her last song. At this point, you should be able to build apps with the basic UI scaffolding and ship them to the App Store. That can get you pretty far, but we've barely scratched the surface on all of the different APIs and frameworks at our disposal. Where should you go now?

First you might want to check out some other iOS books that cover the SDK in greater depth. Cookbook-esque books like *iOS Recipes [WD11]* dedicate chapters to each framework and give you a working knowledge of everything from using the camera to grabbing GPS coordinates. Apple's developer library (https://developer.apple.com/) is the canonical source of iOS information but can be a bit dry and is more useful as a pure API reference. Regardless of the source, almost all of the original Objective-C code should be easily portable to RubyMotion.

You should also dive into the RubyMotion community. The RubyMotion Developer Center (www.rubymotion.com/developer-center) has up-to-date documentation on RubyMotion and goes into more detail on the technical aspects of how it works. A lot of exciting work is being done in taking the original Objective-C APIs and rewriting them in idiomatic Ruby. We used BubbleWrap a few times in this book, which provides great wrappers for more complex tasks such as capturing photos and playing videos. The RubyMotion user group (https://groups.google.com/forum/?fromgroups#!forum/rubymotion) and RubyMotion-Wrappers (http://rubymotion-wrappers.com) are both good resources to learn about new RubyMotion-exclusive frameworks.

Finally, share your work! If you ship an app to the store, definitely post on the user group or send a tweet to the @RubyMotion Twitter account. Or if you write some cool code that makes your life easier, think about making it available to other developers as a gem.

I hope you enjoyed reading this little book. iOS development, RubyMotion or not, is an exciting and evolving field with a vibrant community. I can't wait to see what you come up with!

Bibliography

[Bec02]   Kent Beck. *Test Driven Development: By Example.* Addison-Wesley, Reading, MA, 2002.

[Pin06]   Chris Pine. *Learn to Program.* The Pragmatic Bookshelf, Raleigh, NC and Dallas, TX, 2006.

[WD11]   Paul Warren and Matt Drance. *iOS Recipes: Tips and Tricks for Awesome iPhone and iPad Apps.* The Pragmatic Bookshelf, Raleigh, NC and Dallas, TX, 2011.

More on Ruby

Dig deeper into metaprogramming and create better command-line apps.

Write powerful Ruby code that is easy to maintain and change. With metaprogramming, you can produce elegant, clean, and beautiful programs. Once the domain of expert Rubyists, metaprogramming is now accessible to programmers of all levels. This thoroughly revised and updated second edition of the bestselling *Metaprogramming Ruby* explains metaprogramming in a down-to-earth style and arms you with a practical toolbox that will help you write your best Ruby code ever.

Paolo Perrotta
(250 pages) ISBN: 9781941222126. $38
http://pragprog.com/titles/ppmetr2

Speak directly to your system. With its simple commands, flags, and parameters, a well-formed command-line application is the quickest way to automate a backup, a build, or a deployment and simplify your life. With this book, you'll learn specific ways to write command-line applications that are easy to use, deploy, and maintain, using a set of clear best practices and the Ruby programming language. This book is designed to make *any* programmer or system administrator more productive in their job. This is updated for Ruby 2.

David Copeland
(224 pages) ISBN: 9781937785758. $30
http://pragprog.com/titles/dccar2

The Pragmatic Bookshelf

The Pragmatic Bookshelf features books written by developers for developers. The titles continue the well-known Pragmatic Programmer style and continue to garner awards and rave reviews. As development gets more and more difficult, the Pragmatic Programmers will be there with more titles and products to help you stay on top of your game.

Visit Us Online

This Book's Home Page
http://pragprog.com/book/carubym
Source code from this book, errata, and other resources. Come give us feedback, too!

Register for Updates
http://pragprog.com/updates
Be notified when updates and new books become available.

Join the Community
http://pragprog.com/community
Read our weblogs, join our online discussions, participate in our mailing list, interact with our wiki, and benefit from the experience of other Pragmatic Programmers.

New and Noteworthy
http://pragprog.com/news
Check out the latest pragmatic developments, new titles and other offerings.

Save on the eBook

Save on the eBook versions of this title. Owning the paper version of this book entitles you to purchase the electronic versions at a terrific discount.

PDFs are great for carrying around on your laptop—they are hyperlinked, have color, and are fully searchable. Most titles are also available for the iPhone and iPod touch, Amazon Kindle, and other popular e-book readers.

Buy now at *http://pragprog.com/coupon*

Contact Us

Online Orders:	*http://pragprog.com/catalog*
Customer Service:	*support@pragprog.com*
International Rights:	*translations@pragprog.com*
Academic Use:	*academic@pragprog.com*
Write for Us:	*http://pragprog.com/write-for-us*
Or Call:	+1 800-699-7764

9 781937 785284